Computational
Chemistry
Using the PC

Computational Chemistry Using the PC

Donald W. Rogers

Chemistry Department
Long Island University

New York   Weinheim   Basel   Cambridge

Donald W. Rogers
Chemistry Department
Long Island University
Brooklyn, NY 11201

Library of Congress Cataloging-in-Publication Data

Rogers, Donald, 1932–
    Computational chemistry using the pc / by D. W. Rogers
        p.   cm.
    Includes bibliographical references.
    ISBN 0-89573-770-1
    1. Chemistry—Data processing.   2. Chemistry—Mathematics.
    I. Title.
    QD39.3.E46R64   1990
    542'.85—dc20   90-12309
                                                   CIP

British Library Cataloguing in Publication Data

Rogers, Donald W.
    Computational chemistry using the pc.
    1. Chemistry. Applications of microcomputer systems
    I. Title
    542.8

    ISBN 0-89573-770-1 U.S.

Printed in the United States of America
ISBN 0-89573-770-1   VCH Publishers, Inc.
ISBN 3-52727-937-7   VCH Verlagsgesellschaft

Printing History:
10 9 8 7 6 5 4 3 2 1

Published jointly by:

VCH Publishers, Inc.   VCH Verlagsgesellschaft mbH   VCH Publishers (UK) Ltd.
220 East 23rd Street   P.O. Box 10 11 61   8 Wellington Court
Suite 909   D-6940 Weinheim   Cambridge CB1 1HW
New York, New York 10010   Federal Republic of Germany   United Kingdom

Contents

Preface

This book is an introduction to computational chemistry, molecular mechanics, and molecular orbital calculations, using a personal microcomputer. No special computational skills are assumed of the reader aside from the ability to read and write a simple program in BASIC. No mathematical training beyond calculus is assumed. A few elements of matrix algebra are introduced in Chapter 3 and used throughout.

The treatment is at the upperclass undergraduate or beginning graduate level. Considerable introductory material and material on computational methods is given so as to make the book suitable for self-study by professionals outside the classroom. An effort has been made to avoid logical gaps so that the presentation can be understood without the aid of an instructor. Forty-six self-contained computer projects are included.

The book divides itself quite naturally into two parts: The first six chapters are on general scientific computing applications and the last seven chapters are devoted to molecular orbital calculations, molecular mechanics, and molecular graphics. The reader who wishes only a tool box of computational methods will find it in the first part. Those skilled in numerical methods might read only the second. The book is intended, however, as an entity, with many connections between the two parts, showing how chapters on molecular orbital theory depend on computational techniques developed earlier.

Use of special or expensive microcomputers has been avoided. All programs presented have been run on a 8086-based machine with 640K memory and a math coprocessor. A quite respectable academic program in chemical microcomputing can be started for about $1000 per student. The individual or school with more expensive hardware will find that the programs described here run faster and that more visually pleasing graphics can be produced, but that the results and principles involved are the same. Gains in computing speed and convenience will be made as the technology advances. Even now, run times on an 80386-based machine approach those of a heavily used, time-shared mainframe.

Sources for all program packages used in the book are given in an appendix. All of the early programs (Chapters 1 through 7) were written by the author and are available on a single diskette included with the book. Programs HMO and SCF were adapted and modified by the author from programs in FORTRAN II by Greenwood (*Computational Methods for Quantum Organic Chemistry*, Wiley Interscience, New York, 1972). The more elaborate programs in Chapters 10 through 13 are available at moderate price from Quantum Chemistry Program Exchange, Serena Software, Cambridge Analytical Laboratories and other software sources [see Appendix].

I wish to thank Dr. A. Greenberg of Rutgers University, Dr. S. Topiol of Burlex Industries, and Dr. A. Zavitsas of Long Island University for reading the entire manuscript and offering many helpful comments and criticisms. I wish to acknowledge Long Island University for support of this work through a grant of released time and the National Science Foundation for microcomputers bought under grant #CSI 870827.

Several chapters in this book are based on articles that appeared in *American Laboratory* from 1981 to 1988. I wish to acknowledge my coauthors of these papers, F. J. Mc Lafferty, W. Gratzer, and B. P. Angelis. I wish to thank the editors of *American Laboratory*, especially Brian Howard, for permission to quote extensively from those articles.

Live joyfully with the wife whom thou lovest all the days of the life of thy vanity, which He hath given thee under the sun, all the days of thy vanity for that is thy portion in this life, and in thy labor which thou takest under the sun.

Ecclesiastes 9:9

This book is dedicated to Kay

1 *Iterative Methods*

One of the most important methods of modern computation is that of solution by iteration. The method was well known long ago but only came into widespread use with the speed of the modern computer. Normally, one uses iterative methods when ordinary (analytical) mathematical methods fail or are too time-consuming to be practical. Even relatively simple mathematical procedures, for example, integration by parts, may be time-consuming because of extensive algebraic manipulation.

A common iterative procedure is to solve the problem of interest by repeated calculations, often of a very elementary nature, that do not initially give the correct solution, but that get closer and closer to it as the calculation is repeated, perhaps very many times. The approximate solution is said to *converge* on the correct solution. Although no human would be willing to repeat a calculation many thousands of times to obtain one answer, the computer does, often in a reasonable time period.

AN ITERATIVE ALGORITHM

The first illustrative problem comes from quantum mechanics. A problem in *radiation density* can be set up but not solved by conventional means. Faced with an equation that is not soluble by ordinary methods, one guesses a solution and substitutes it into the equation, applying a test to see if the equation is satisfied. Of course, it is not because the "solution" was only a guess, but an iterative program can be set up so as to modify the guess to make it closer to the right answer. The same test is applied again and fails again; the routine is iterated until the modified guess converges on the right answer. Eventually, the test shows that the guess has been modified until it satisfies the equation, i.e., until it has been transformed into the solution of the equation.

With this problem, many questions present themselves immediately: How good does a guess have to be? How do we know that the modification procedure makes the guess better, not worse? How long (how many iterations) will it take? How do we know when to stop? These questions

1

and others like them will play an important role in this book because they are central to many computational methods in chemistry. You will not be surprised to learn that the answers to questions like these vary from one problem to another and cannot be set down once and for all.

Before solving problems in radiation density, we shall solve an introductory problem concerning density variation over a one-dimensional x space.

Space

Any variable measured along an axis defines a space. For example, plotting x along the horizontal axis defines a one-dimensional x space. Space in the x, y, and z dimensions is often called a cartesian space in honor of the mathematician René Descartes. If probability is plotted as an axis, one has a probability space; if velocity is plotted, one has a velocity space, and so on.

Density

Density, mass per unit volume (V), depends on the number of particles of matter in a defined space, for example, the atoms of helium per cubic centimeter. Though it is customary to describe g grams of a gas in terms of mass density g/V, a number density, (number of atoms)$/V$, would give the same information, because we know the mass of each particle.

EXERCISE 1-1

A 1-m metal rod, 1 cm^2 in cross section is made of an alloy of variable composition such that its mass density ρ increases over its length at the rate of 0.050 g cm^{-3}/cm along the length of the rod. (a) What is the mass of the rod if its density at the light end is 2.00 g cm^{-3}? (b) Where is its balance point (center of mass)?

Solution 1-1 Although the rod has volume, we are considering variations only in the x dimension; hence, the problem is set in a one-dimensional x space. Let $x = 0$ at the light end of the rod. If we plot ρ vs. x, we have a two-dimensional (ρ, x) density-x space.

The rod increases in density at a continuous rate of 0.050 g cm^{-4} over its length and has a density of 2.00 g cm^{-3} at its light end; hence, $\rho = 2.00 + 0.050x$ at any point. The cross section is 1.00 cm^2; hence an increase of dx is also an increase of dV cm^3 in volume of metal. The weight of an incremental section is $dw =$

$\rho \, dV = \rho \, dx$. With $l = 100$ cm, the total weight of the bar is

$$w = \int_0^{x=l} dw = \int_0^l \rho \, dx = \int_0^l (2.00 + 0.050x) \, dx$$

$$= 2.00x + (0.050x^2)/2 \Big|_0^{l=100} = 450 \text{ g}$$

For the balance point, we wish to find the value of b for which the integral

$$\int_0^b (2.00 + 0.050x) \, dx = 450/2$$

that is,

$$2.00b + (0.050b^2)/2 = 225$$

$$b^2 + 80b = 9000$$

$$b = 63 \text{ cm}$$

Radiation Density

If we think in terms of the particulate nature of light (wave–particle duality), the number of particles of light (photons) in a unit of frequency space constitutes a number density. The black-body radiation curve [Fig. 1-1(a)] is a plot of radiation density, on the vertical axis, as a function of frequency ν, plotted on the horizontal axis. The ρ and ν axes are the coordinates of a two-dimensional density–frequency space. The radiation density is the probable number of particles in an infinitesimal frequency interval ν to $\nu + d\nu$. Radiation density is a function of frequency and temperature, $\rho(\nu, T)$. Because frequency times wavelength λ is the velocity of light $\nu\lambda = c = 2.99793 \times 10^8$ m s^{-1}, an equivalent functional relationship exists between density of radiation and wavelength. The radiation density can be written in an equivalent form, $\rho(\lambda, T)$. The *intensity* I of electromagnetic radiation within any narrow wavelength interval is directly proportional to the number density of light particles or photons. It is also directly proportional to the power output of a light sensor or photomultiplier; hence, both I and ρ are measurable quantities.

Wien's Law

In the nineteenth century, Wien observed experimentally that the maximum of the black-body radiation curve $\lambda_{\max}$ shifts with the temperature according to the equation

$$\lambda_{\max} T = 2.90 \times 10^{-3} \text{ m K} \tag{1-1}$$

where λ is in meters and T is the absolute temperature in kelvins.

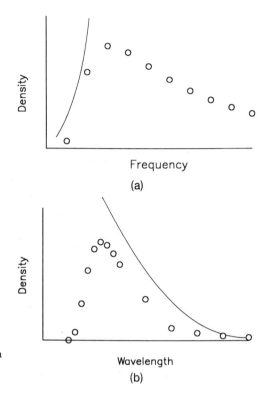

FIGURE 1-1 Density of black-body radiation **(a)** as a function of ν and **(b)** as a function of λ.

The classical expression for radiation,

$$\rho(\nu, T)\, d\nu = \{8\pi kT/c^3\}\nu^2\, d\nu$$

must fail to express these curves because $\rho = (\text{const})\nu^2$ is a parabola [solid line, Fig. 1-1(a)] and does not even have a relative maximum. Max Planck showed that the quantization of energy in units of $\Delta E = h\nu$ leads to

$$\rho(\nu, T)\, d\nu = \{8\pi h\nu^3/c^3\}\, d\nu/[\exp(h\nu/kT) - 1] \qquad (1\text{-}2)$$

where h is Planck's constant and k is Boltzmann's constant. This can be transformed to

$$\rho(\lambda, T)\, d\lambda = \{8\pi hc)/\lambda^5\}\, d\lambda/[\exp(hc/\lambda kT) - 1] \qquad (1\text{-}3)$$

EXERCISE 1-2

Carry out the transformation of Eq. (1-2) to Eq. (1-3).

By setting $d\rho/d\lambda = 0$, one can differentiate and show that the equation

$$e^{-x} + x/5 = 1 \qquad (1\text{-}4)$$

holds at the maximum of Fig. 1-1(b) where

$$x = hc/\lambda kT \qquad (1\text{-}5)$$

EXERCISE 1-3

Obtain Eq. (1-4) from Eq. (1-3).

Coding

Translating an algorithm or procedure into an ordered series of statements in a computer language—a program—and entering the program into a computer is called coding. BASIC and FORTRAN are the two languages we shall use in this book, but you need spend little time coding. You will be asked to code a few of the programs that follow for practice, but most programs are available on disk, already coded. You will, however, need to learn enough of the language to control your program: to input information when necessary and to understand the output format of the results.

Do not work from your original disk. Work from a copy so that the originals will not be lost in the event of a blunder or machine crash.

The Programs

Most of the programs referred to in Chapters 1 through 9 are on the disk provided with this book. They can be loaded into computer memory and then modified as desired by retyping one or more BASIC statements. Many programs require modification in the DATA block. Modified programs can be saved to a unique file name or discarded by typing NEW. Any program saved to a file name of one of the existing programs will erase (overwrite) that program. Some systems require the suffix as part of the file name, e.g., LOAD"WIEN.BAS". Most systems do not require capital letters; that is an old convention that we shall use to make file names and commands easier to distinguish from text.

COMPUTER PROJECT 1-1 | Wien's Law

The first project is devoted to solving Eq. (1-4) for x iteratively. When x has been determined, the remaining constants can be substituted into

$$\lambda T = hc/kx \qquad (1\text{-}6)$$

where h is Planck's constant, c is the velocity of light and k is the

Boltzmann constant. The result is a test of agreement between Planck's quantum hypothesis and Wien's displacement law [Eq. (1-1)].

Procedure One approach to the problem is to select a value for x that is obviously too small and to increment it until the equation is satisfied. This is the method of program WIEN, where the initial value of x is taken as 1 (clearly, $e^{-1} - \frac{1}{5} < 1$). Program WIEN is written in both extended and condensed format (condensed as WIEN1). Condensation using a colon (:) between statements on the same line saves space.

(a) Vary the size of the increment to x in program WIEN, statement 20. Tabulate the increment size, the computed result for x, the calculated Wien constant, and the run time as estimated using an ordinary watch. Comment on the relationship among the quantities tabulated.

(b) Change Eq. (1-4) so that the second term on the left is x. Solve for this new equation, Eq. (1-4.1). Solve for Eq. (1-4.2) with a second term of $x/2$, Eq. (1-4.3) with $x/3$, and Eq. (1-4.4) with $x/4$ etc. Tabulate the values for x. Is x a sensitive function of the denominator of the second term?

(c) Devise and discuss a scheme for more efficient convergence. For example, some scheme that uses large increments for x when x is far from its convergence value and small values for the increment when x is near its true value would be more efficient than the preceding schemes. How, in more detail, could this be done. Try coding your scheme.

COMPUTER PROJECT 1-2 | *The van der Waals Cubic*

The van der Waals equation for 1 mol of a real gas is

$$(p + a/V_m^2)(V_m - b) = RT \qquad (1\text{-}7)$$

where V_m is the molar volume of a gas and a and b are the van der Waals constants (Atkins, 1986). This form can be rearranged to give a cubic in V_m:

$$V_m^3 - (b + RT/p)V_m^2 + (a/p)V_m - (ab/p) = 0 \qquad (1\text{-}8)$$

Procedure Determine the molar volume of nitrogen at 500 K and 100 atm pressure using the van der Waals equation. The van der Waals constants for N_2 are

$$a = 1.390 \text{ dm}^6 \text{ atm mol}^{-2}$$
$$b = 3.913 \times 10^{-2} \text{ dm}^3 \text{ mol}^{-1}$$

Substituting into the equation of interest,

$$V_m^3 - 0.4495V_m^2 + 1.390 \times 10^{-2}V_m - 5.439 \times 10^{-4} = 0 \quad (1\text{-}9)$$

The cubic can be solved analytically, but, for the purpose of this and the next exercise, we shall solve it iteratively, using program VAN.

We need an initial guess for V_m. An approximate solution using the ideal gas law gives $V_m = 0.4105$ under the conditions cited. An initial value of 0.1 should be much smaller than the true value; hence, it serves as a starting point for an incremental calculation.

(a) Modify the program from Computer Project 1-2 so as to accept a, b, p, and T as input values. This modification makes the program general for gases other than nitrogen. Use the new program to determine the molar volume for carbon dioxide, krypton, and propane at various pressures and temperatures. van der Waals constants are available in most general references and handbooks of chemistry.

COMPUTER PROJECT 1-3 | *Roots of the Secular Matrix*

In Chapter 9, we shall encounter a matrix equation

$$\begin{bmatrix} 210 - 42x & 42 - 9x \\ 42 - 9x & 12 - 2x \end{bmatrix} = 0 \qquad (1\text{-}10)$$

Although it is not the method we shall prefer in this book, one way of solving this matrix equation is to expand it as a determinant. To do this, multiply the binomials at top left and bottom right and then, from this product, subtract the product of the remaining two binomials. This difference is set equal to 0:

$$(210 - 42x)(12 - 2x) - (42 - 9x)^2 = 0 \qquad (1\text{-}11)$$

This equation is a quadratic and has two roots. For quantum-mechanical reasons, we are interested in the lower root. By inspection, $x = 0$ leads to a large number on the right of Eq. (1-11). Letting $x = 1$ leads to a smaller number on the right of Eq. (1-11) but one that is still greater than 0. Evidently increasing x approaches a solution of Eq. (1-11), that is, a value of x for which both sides are equal. By systematically increasing the value of x beyond 1, we will approach one of the roots of the secular matrix.

The five-line program (program ROOT on the disk),

```
10   X = 0
20   X = X + 1
30   A = (210 − 42*X)*(12 − 2*X) − (42 − 9*X)**2
40   IF A > 0 GOTO 20
50   PRINT X:END
```

increments X by 1 on each iteration. It prints out 5 when the polynomial on the right is less than 0, i.e., when we have gone past the root because X is too large. The program did not exit from the loop on $X = 4$, but it did on $X = 5$; hence, X is between 4 and 5. By letting $X = 4$ in the first line and changing the second statement to increment X by 0.1, we get 5 again;

hence, X is between 4.9 and 5.0. Letting $x = 4.9$ with an increment of 0.01 yields 4.94 and so on, until the increment 0.00001 yields the lower root $x = 4.93488$. Six significant digits are needed to complete the calculation in Chapter 9.

Although we shall not need it for the calculation in Chapter 9, we may be curious to evaluate the second root, and we shall certainly want to be sure that the root we have found is the smaller of the two. One of the problems asks you to write a program to evaluate the right side of Eq. (1-11) at integral values between 1 and 100 to make an approximate location of the second root. Another problem asks you to write a second program to locate the second root of matrix Eq. (1-10) to a precision of six digits.

PROBLEMS | Chapter 1

1 Determine the value of x in Exercise 1-1 by solving the quadratic

$$x^2 + 80x - 9000 = 0$$

iteratively. Use a modification of the program in Computer Project 1-2.

2 Suppose the metal bar in Exercise 1-1 is of a uniform density but is shaped in the form of a right triangle with a base of 1.00 m and a height of 2.00 cm. Where is its balance point?

3 The energy of radiation at a given temperature is the integral of radiation density over all frequencies

$$E = \int_0^\infty \rho(\nu, T) \, d\nu$$

Find E from the known integral

$$\int_0^\infty \frac{x^3 \, dx}{\exp(x) - 1} = \frac{\pi^4}{15}$$

and compare the result with the Stefan–Boltzmann law

$$E = (4\sigma/c)T^4$$

where c is the velocity of light and σ is an empirical constant equal to 5.6697×10^{-8} J m^{-2} s^{-1}. The value of the "known integral" is not obvious; we shall determine it numerically in the next chapter.

4 Analysis of the electromagnetic radiation spectrum emanating from the star Sirius shows that $\lambda_{max} = 260$ nm. Estimate the surface temperature of Sirius.

5 Obtain the cubic form in Computer Project 1-2 from the conventional form of the van der Waals equation.

6 Cite a fundamental objection to the method for solving cubics given in Computer Project 1-2.

7 Incorporate the program fragment given in Computer Project 1-3 into a larger program that locates the root to the desired six-digit precision in one run, i.e., one that does not have to be repeated for each digit of precision.

8 Write a program to determine 100 values of the function $f(x)$ in Eq. (1-11). Determine the location of the higher root.

9 Using the method of Computer Project 1-3, modify program ROOT so as to determine the value of the higher root of Eq. (1-11) to an accuracy of six digits. (*Ans.* 51.0647.) Store the modified program as ROOT1.

REFERENCES

Atkins, P. W., 1986. *Physical Chemistry*, 3rd ed. W. H. Freeman, New York.

Mc Quarrie, D. A., 1983. *Quantum Chemistry*. University Science Books, Mill Valley, CA.

2 | *Numerical Integration*

The term *quadrature* was used by early mathematicians to mean finding a square with an area equal to the area of some geometric figure other than a square. It is used in numerical integration to indicate the process of summing the areas of some number of simple geometric figures to approximate the area under some curve, i.e., to approximate the integral of a function.

This discussion will be limited to functions of one variable that can be plotted on a two-dimensional surface over the interval considered and that constitute the upper boundary of a well-defined area. The functions selected for illustration are simple and well-behaved; they are smooth, single-valued, and have no discontinuities. When discontinuities or singularities do occur, we shall integrate up to the singularity but not include it.

Contrary to the impression that one might have from a traditional course in introductory calculus, well-behaved functions that cannot be integrated in closed form are not mathematical curiosities; they are more common than those that can be integrated. Examples are the gaussian or standard error function and the related function that gives the distribution of molecular or atomic speeds in spherical polar coordinates. The famous black-body radiation curve (Fig. 1-1), which inspired Planck's quantum hypothesis, is not integrable in closed form over an arbitrary interval.

Heretofore, the integral of a function of this kind was usually approximated by expressing it as an infinite series and evaluating some arbitrarily limited number of terms of the series. This always leads to a truncation error that depends on the number of terms retained in the sum before truncation.

Numerical integration may be used instead of series solution or when the analytical form of the function is not known because the functional relationship exists as an instrument plot or a collection of paired measurements. This is the common case for data that have been obtained in an experimental setting. An example is the function describing a chromatographic peak that may or may not approximate a gaussian function.

Several related "rules" or algorithms for numerical integration will be described and examples of their application will be given both for functions that exist in analytical form and are integrable in closed form (for analysis of the method by comparison) and for functions that exist in analytical form but are not integrable. We shall use the term *analytical form* to indicate a closed algebraic expression such as

$$y = x^2 \tag{2-1}$$

as contrasted with functions that have only an infinite series, e.g.,

$$C_p = a + bT + cT^2 + \cdots \tag{2-2}$$

as their algebraic form. Equation (2-1) is an analytical form that has a closed integral. The gaussian function

$$f(x) = (2\pi)^{-1/2} \exp\left(\frac{-z^2}{2}\right) \tag{2-3}$$

is a closed analytical form but it has no closed integral. (Try to integrate it!)

ITERATIVE QUADRATURE ALGORITHMS

The Rectangular Rule

Suppose that one can graph an arbitrary function $f(x)$ over an interval a to b, and that the function is well-behaved on this interval. We can estimate the integral $\int f(x)\,dx$ by selecting an arbitrary point on the curve, for example point a, and supposing that the entire curve is horizontal over the interval with height $f(a)$ [Fig. 2-1(a)]. A rectangle bounded by the verticals at a and b, the x axis, and the horizontal through $f(a)$ is the first and most crude approximation to the integral of $f(x)$. It is also closest to the classical meaning already given for the term quadrature.

The estimate obtained by this method is poor and it is worse for steeper or more complicated functions. Nevertheless, it serves as the basis of a method that can be used to surprising advantage, given the speed and accuracy of today's computers. We will break the interval $[a, b]$ into n subintervals of equal width w, so that $a = x_0$, $b = x_n$, and $x_i = a + iw$, where $i = 0, 1, 2, \ldots, n - 1$. The areas of the regions bounded horizontally by $x = x_i$, $x = x_i + iw$, $y = f(x_i)$, and the x axis are shown in Fig. 2-1(b). The estimation method is the same as before but $f(x_i)$ replaces $f(a)$ in each of the subintervals and the integral is approximated by a sum of areas. The reader can see that the error in the approximation decreases with an increase in the number of subintervals taken. Therefore, we can make our integration method more accurate without limit except for

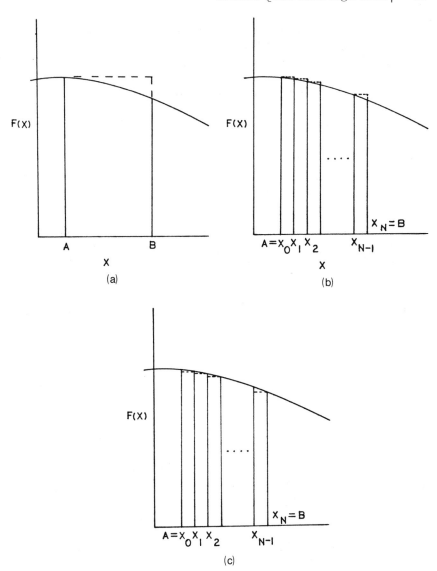

FIGURE 2-1 The integral $\int F(x)\,dx$ over the interval $[a, b]$ approximated by the area of **(a)** the rectangle $(F(a))(b - a)$, **(b)** the sum of rectangles $F(x_i)w$, and **(c)** the sum of rectangles $F(x_{i+1})w$, where $i = 0, 1, 2, \ldots, N - 1$ and w is a subinterval $(b - a)/N$.

practical or machine limits. This simplest form of quadrature is called *rectangular rule* integration (Scheid, 1968).

The formula for rectangular rule integration is

$$\int_a^b f(x)\, dx = f(x_0)w + f(x_1)w + f(x_2)w + \cdots + f(x_{n-1})w \quad (2\text{-}4)$$

Because w appears in each term, it can be factored out:

$$\int_a^b f(x)\, dx = w(f(x_0) + f(x_1) + f(x_2) + \cdots + f(x_{n-1}))$$

$$= w \sum_0^{n-1} f(x_i) \quad (2\text{-}5)$$

As seen in Fig. 2-1(c), we could have chosen our heights differently, from right to left starting with $f(b)$, which would give a different area estimation. The estimation formula for this approach is very similar to the preceding one,

$$\int_a^b f(x)\, dx = f(x_1)w + f(x_2)w + \cdots + f(x_{n-1})w + f(x_n)w \quad (2\text{-}6)$$

or

$$\int_a^b f(x)\, dx = w(f(x_1) + f(x_2) + \cdots + f(x_n))$$

$$= w \sum_1^n f(x_i) \quad (2\text{-}7)$$

but the summed area is smaller than the integral and tends toward it as the number of rectangles is increased.

For many functions of scientific interest, $f(x)$ decreases or increases *monotonically* (without changing the sign of the slope) over the subinterval. Hence, $f(x)$ at either the leftmost or the rightmost extreme of the subinterval is the worst approximation to a representative or average $f(x)$ on the subinterval. A better estimation of the true area under the function can be obtained by multiplying the width of the subinterval into the value of $f(x)$ at some point between the beginning and endpoints of the subinterval. A natural point to choose is the midpoint, which leads to a variant of the rectangular rule known as the *midpoint rule*:

$$\int_a^b f(x)\, dx = w \sum_0^{n-1} f(x_i + \tfrac{1}{2}w) \quad (2\text{-}8)$$

The Trapezoidal Rule

Another approach to minimizing the errors due to the rather crude assumptions of the rectangular rule is to approximate the area of a

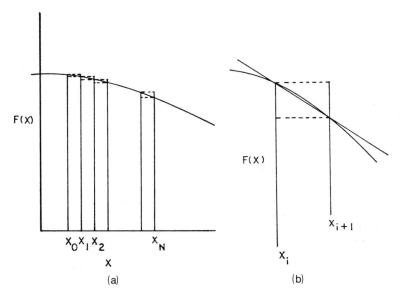

FIGURE 2-2 The integral $\int F(x)\,dx$ approximated by (a) averaging methods shown in Fig. 2-1(b) and (c). Fig. 2-2(b): the trapezoidal rule.

subinterval of the function by averaging two rectangular rule quadratures as shown in Fig. 2-2(a) and (b). This average method has the popular name *trapezoidal rule* integration (Scheid, 1968).

The trapezoidal rule formula is easily found by averaging the two alternative rectangular rule formulas

$$\int_a^b f(x)\,dx = \tfrac{1}{2}w\big(f(x_0) + 2f(x_1) + 2f(x_2) + \cdots$$

$$+2f(x_{n-1}) + f(x_n)\big) \qquad (2\text{-}9)$$

$$= \tfrac{1}{2}w\left(f(x_0) + \sum_{1}^{n-1} 2f(x_i) + f(x_n)\right)$$

where n is the number of subintervals.

Simpson's Rule

In the trapezoidal rule we used the fact that two points determine a straight line to obtain the cover of each subinterval. We will now demand more information, three functional values, to improve the approximation.

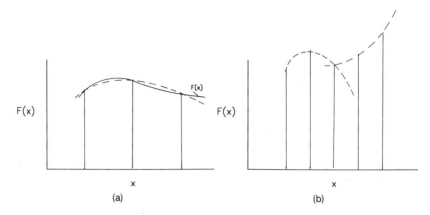

FIGURE 2-3 Approximating an integral by Simpson's rule: (**a**) Two subintervals approximated by a parabolic arc and (**b**) four subintervals approximated by two parabolic arcs.

The interval a to b is partitioned into an *even* number of subintervals and three consecutive points are used to determine the unique parabola that is the covering for the first subinterval pair; see Fig. 2-3(a). The area under this parabolic arc is $\frac{1}{3}w(f(x_i) + 4f(x_{i+1}) + f(x_{i+2}))$. Summing for successive subinterval pairs over the entire interval, the method known as Simpson's rule (Scheid, 1968) is found; see Fig. 2-3(b). Looking at the following formula, it becomes obvious that a simple loop will implement it on a computer:

$$\int_a^b f(x)\,dx = \frac{1}{3}w(f(x_0) + 4f(x_1) + 2f(x_2) + 4f(x_3) + \cdots$$

$$+ 2f(x_{n-2}) + 4f(x_{n-1}) + f(x_n))\quad (2\text{-}10)$$

EXERCISE 2-1

Show that the area under a parabolic arc that is convex upward is $\frac{1}{3}w(f(x_i) + 4f(x_{i+1}) + f(x_{i+2}))$, where w is the width of a subinterval $x_{i+1} - x_i$.

Solution 2-1 The area under a parabola (see computer projects that follow) is $\frac{2}{3}(bh)$ where b is the base of the figure and h is its height. The areas of parts of the figure diagrammed for Simpson's rule integration are shown on the figure.

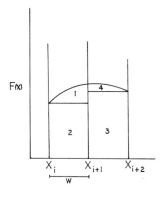

$$A = \tfrac{2}{3}w\big(f(x_{i+1}) - f(x_i)\big) + wf(x_i) + wf(x_{i+2})$$
$$+ \tfrac{2}{3}w\big(f(x_{i+1}) - f(x_{i+2})\big)$$
$$= w\big(\tfrac{2}{3}f(x_{i+1}) + \tfrac{1}{3}f(x_i) + \tfrac{2}{3}f(x_{i+1}) + \tfrac{1}{3}f(x_{i+2})\big)$$
$$= \tfrac{1}{3}w\big(f(x_i) + 4f(x_{i+1}) + f(x_{i+2})\big)$$

EFFICIENCY AND MACHINE CONSIDERATIONS

We have selected a simple test function for some of the integration schemes previously given. The function $f(x) = 100 - x^2$ is a smooth, monotonically decreasing parabolic curve over the interval $0 \le x \le 10$. It has a closed definite integral over this interval of 666.666 units as seen in Fig. 2-4. The function is well-behaved and integration is easy by the rules given here over the first half of the interval, but not so easy over the second half of the interval owing to its increasing steepness. (Note that

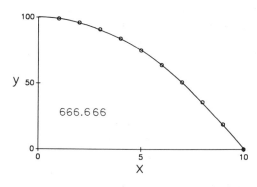

FIGURE 2-4 Areas in the positive quadrant under $f(x) = 100 - x^2$ over the interval $[0, 10]$.

TABLE -1 Iterative Approach to the Closed Integral for Rectangular Rule Integration of $f(x) = 100 - x^2$

Iterations	Area	Approx. Time, s[a]	Error Percent
1	1000	< 1	50
10	715	< 1	7
100	671.65	< 2	0.7
1000	667.16	< 14	0.07
10000	666.729	~ 230	9×10^{-3}

[a]Run times should be regarded as relative because they depend on both the computer used and the software.

steep functions can be integrated by an algorithm that sums horizontal slices of the area under the curve rather than vertical ones.)

Applying the rectangular rule for quadrature, we divide the interval into n subintervals and follow Eq. (2-1) to approximate the area under $f(x)$ on the interval $[a, b]$. Some combination of the BASIC statements W = (B − A)/N and S = S + (100 − X*X)*W will do, along with a mechanism for advancing X as it is moved along the interval in equal increments. Program INTEG shows one way of doing this. Alternatively, $f(x)$ can be calculated at the rightmost limit of x in each subinterval and multiplied into w yielding a sum that approaches the integral from the low side.

Neither rectangular rule algorithm is particularly efficient. Table 2-1 shows the approach of rectangular rule integration of the test function on the known closed value for 1, 10, 100, 1000, and 10,000 iterations of Eq. (2-4) or Eq. (2-6). Run time on an 8-MHz microcomputer using interpreted BASIC is about 13 s for 1000 iterations and about 4 min for 10,000 iterations using Eq. (2-4).

Equation (2-5), though mathematically equivalent to Eq. (2-4), implies a saving of machine time by taking the summation of $f(x_i)$ separately from the multiplication by w. That is, by Eq. (2-4), multiplication of $f(x_i)$ by w is in the summation loop and occurs many times; by the algorithm implied by Eq. (2-5), the summation is carried out first and multiplication of the sum by w is carried out once. Run-time reduction is small, however, amounting to only 2% or so.

Table 2-1 shows that the approximation to the closed integral improves as the number of iterations increases, continuing the trend started when the number of iterations was small and the approximation crude. One is tempted to think of the approximation as getting better without limit, the sum approaching the integral as Achilles approached the tortoise. This does not occur because of machine rounding error, because

TABLE 2-2 Comparison of Four Common Algorithms for
Numerical Integration of $f(x) = 100 - x^2$

Iterations	Rectangular	Trapezoidal	Midpoint	Simpson
2	875	625	687.5	666.6666
4	781.25	656.25	671.875	666.6666
10	715	665	667.5	666.6667
50	676.6	666.6	666.7	666.6665
100	671.6503	666.6503	666.6751	666.6668
150	669.9926	666.6593	666.6705	666.6668
200	669.1627	666.6627	666.669	666.667
300	668.332	666.6653	666.668	666.6672
500	667.6663	666.6663	666.6679	666.6669

computers cannot "carry" an infinite number of significant figures. It is
not difficult to find sums that, at some number of iterations, begin to
diverge from, rather than approach, the integrals they are supposed to
represent (Norris, 1981).

Table 2-2 shows that trapezoidal, midpoint, and Simpson's rule
integrations approach the exact solution of the closed integral much more
rapidly than the rectangular rule, being within 1 or 2 ppm of it after 500
iterations. Simpson's rule is often the most efficient algorithm of these
three in terms of closeness of approach for the fewest iterations (Rice,
1983). In this illustrative integration, Simpson's rule is within a part per
million or so of the closed integral after only two iterations, but this is
hardly a fair test because Simpson's rule is a parabolic approximation and
the test function happens to be a parabola.

COMPUTER PROJECT 2-1 | *Chemical Statistics*

The gaussian distribution for the probability of random events is

$$p(x) = \left[\frac{1}{(2\pi)^{1/2}\sigma} \right] \exp\left(-\frac{(x_i - \mu)^2}{2\sigma^2} \right) \qquad (2\text{-}11)$$

where μ is the arithmetic mean and σ is the population standard
deviation. Equation (2-11) is widely used in experimental chemistry, most
commonly in statistical treatments of experimental uncertainty (Young,
1962).

For convenience, it is common to make the substitution

$$z = \frac{x_i - \mu}{\sigma} \qquad (2\text{-}12)$$

where z is called the *standard normal deviation*. The gaussian function over interval $[a, b]$ becomes

$$P(z) = (2\pi)^{-1/2} \int_a^b e^{-z^2/2} dz \qquad (2\text{-}13)$$

Because of this substitution, distributions having different μ and σ can be compared using the same curve, which is frequently called the *normal curve*.

The integral of the gaussian distribution function does not exist in closed form, but it is a simple matter to calculate the value of $p(z)$ for any value of z; hence, numerical integration is appropriate. The approach of the rectangular rule method on the accepted values (Young, 1962) of this integral is slow. At 100 iterations, the area sum for the integral of the gaussian from 0 to 1 is 0.22% too large. Taking 1000 rectangles yields 0.3414, in 2 s, which is 0.029% larger than the accepted value.

As with the previous test function, the integral of this function is approached in a few iterations by Simpson's rule. Application of Simpson's rule gives four-place accuracy or better at millisecond run time in only six iterations. For many applications in applied probability and statistics, four significant figures are more than can be supported by the data.

The iterative loop for approximating an area can be nested in an outer loop that prints the area under the gaussian error curve for each of many increments in z. If the output is arranged in appropriate rows and columns, a table of areas under one-half the gaussian error curve can be generated, for example, from 0.0 to $4.0z$, resulting in printed values of the area at intervals of $0.01z$. This is suggested to the interested reader as an exercise. A 400-entry table can be generated in about 2 min at 8-MHz clock speed.

The practical value of generating a table of gaussian error function areas is small because many such tables are available in statistics books. The method, however, can be applied to derivative functions of the gaussian error function with only minor modifications, resulting in generation of tables of considerable practical importance (see the following text).

The mean and standard deviation of a population are fundamental statistical parameters because they specify one and only one gaussian distribution through Eq. (2-11). Applications of the area calculation of a gaussian function as, for example, in detection of outliers, estimation of the probability that a continuous variable will fall in a given subinterval of its range, sampling theory, establishing confidence limits, accessing the significance of difference between means, and rejection of data are well documented (Rogers, 1983; Balaam, 1972).

Procedure Suppose that the total serum cholesterol level in normal adults has been established as 200 mg/100 ml (mg%) with a standard deviation of 25 mg%. A patient's serum is analyzed for cholesterol and found to contain 265-mg% total cholesterol.

(a) May we say at the 0.95 confidence level that the patient's cholesterol is abnormally elevated?

(b) May we reach the same conclusion at the 0.99 confidence level?

(c) If a patient is just at the 95% confidence level, there is a 5% probability that his cholesterol is randomly high and not indicative of pathology. What is the probability that the cholesterol reading obtained for *this* patient (265 mg%) owed to chance factors and does not indicate pathology?

Determine z. Run the program INTEG3 from the disk. Use the results to answer questions (a) through (c). Turn in the results of this experiment with a short discussion.

COMPUTER PROJECT 2-2 | *Maxwell–Boltzmann Distribution Laws*

The Maxwell–Boltzmann *distribution function* (Levine, 1983; Kauzmann, 1966) for atoms or molecules of a gaseous sample is

$$F(\mathbf{v}_x) = \left(\frac{m}{2\pi kT}\right)^{1/2} \exp\left(\frac{-m\mathbf{v}_x^2}{2kT}\right) \tag{2-14}$$

for molecular velocity vectors $\mathbf{v}_x$ about their arithmetic mean $\mathbf{v}_x = 0$ along an arbitrarily selected x axis. The distribution function for molecular speeds v is

$$G(v) = \left(\frac{m}{2\pi kT}\right)^{3/2} \exp\left(\frac{-mv^2}{2kT}\right) 4\pi v^2 \tag{2-15}$$

where $v = (v_x^2 + v_y^2 + v_z^2)^{1/2}$. These lead to the familiar distribution curves in Fig. 2-5. The velocity vector (boldface) has magnitude and direction; hence, $\mathbf{v}$ can be negative. Speed is a scalar and is always positive.

The expectation value of $\mathbf{v}_x$ is

$$\langle \mathbf{v}_x \rangle = \int_a^b \mathbf{v}_x f(\mathbf{v}_x)\, d\mathbf{v}_x \tag{2-16}$$

between the limits $a < \mathbf{v}_x < b$. By the nature of the model, if we set the limits of integration at $-\infty$ and ∞, the expectation value is the mean $\langle \mathbf{v}_x \rangle = \bar{\mathbf{v}}_x = 0$. The expectation value of v is

$$\langle v \rangle = \int_a^b v G(v)\, dv \tag{2-17}$$

The expectation value of the speed is its most probable value and can be obtained by numerical integration of Eq. (2-17) from 0 to ∞ or by differentiation of the distribution function (Kauzmann, 1966) to obtain $v_{mp} = (2kT/m)^{1/2}$.

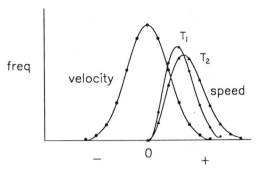

FIGURE 2-5 The frequency distribution function of velocity and speed for a collection of particles in random (gaussian) motion. The speed curves are at temperatures $T_1 < T_2$. The curve for velocity is not on the same scale as the curves for speed.

In chemical kinetics, it is often important to know the proportion of particles with a speed that exceeds a selected speed v'. The probability of finding a particle with a speed from 0 to v' is the integral of the distribution function over that interval,

$$\int_0^{v'} G(v)\, dv = \left(\frac{m}{2\pi kT}\right)^{3/2} \int_0^{v'} \exp\left(\frac{-mv^2}{2kT}\right) 4\pi v^2\, dv \quad (2\text{-}18)$$

The probability of finding a particle with a molecular speed somewhere between 0 and ∞ is 1.0 because negative molecular speeds (in contrast to velocities) are impossible; hence, the probability of speeds in *excess* of v' is $1.0 - \int_0^{v'} G(v)\, dv$.

It is convenient to reason in terms of the proportion of particles having a velocity in excess of v_{mp}. The most probable speed works as a normalizing factor, permitting us to generate one curve that pertains to gases in general rather than a different curve for each molecular weight and temperature. The integral of $G(v)\, dv$, however, cannot be obtained in closed form. The function is usually integrated by parts (Levine, 1983) with the use of a scaling factor to yield a three-term equation that is evaluated to give the appropriate ratio of particles with speeds in excess of v'/v_{mp} as a function of v'/v_{mp}. The calculated curve (which passes near the point $f = 0.5$ at $v' = v_{mp}$) can be made specific to any gas at any temperature from a knowledge of v_{mp}. This technique does not really escape the problem of nonintegrable functions, however, because the second term in the evaluation for the frequency factor is a gaussian.

It is also possible to integrate Eq. (2-18) directly by numerical means and to subtract the result from 1.0 to obtain the proportion of particles with speeds in excess of v'/v_{mp}.

Procedure Using Program KAUZ, draw the curve of the fraction of particles with speeds in excess of v'/v_{mp} as a function of v'/v_{mp}. Compare your results with the literature values (Kauzmann, 1966; Rogers and Gratzer, 1984). One can INPUT U, the upper limit of velocity v', and carry out a numerical integration from $A = 0$ to $B = U$ to obtain the area under the gaussian from 0 to v'/v_{mp}. Using such a scheme, the illustrative table in Kauzmann was generated in about 4 s using interpreted BASIC at 8-MHz clock speed.

Because the computer cannot store an infinite number of bits, computations leading to very small and very large numbers are often inaccurate. Results of the present calculation are poor at very high velocities. Fortunately, an approximation formula for $v'/v_{mp} \gg 1.0$ is known (Kauzmann, 1966):

$$f(v) = \left(\frac{1}{\pi^{1/2}}\right)\exp(-v^2)\left(2v + \frac{1}{v}\right) \qquad (2\text{-}19)$$

for the fraction of molecular velocities that are substantially in excess of v_{mp}. Particles moving with these extreme velocities are rare but important because, in many reactions, only very fast-moving molecules have sufficient energy to react. The proportion of very energetic molecules, say those with speeds in excess of $4v_{mp}$ relative to the total number of molecules, increases rapidly with temperature; this is the cause of the exponential rise of the reaction rate with temperature observed in many reactions (Arrhenius' rate equation).

Find the speed below which 75% of N_2 molecules move at 500 K. Why is f near but not equal to 0.5 when $v' = v_{mp}$? Use Program KAUZ to determine the ratio of the median speed to v_{mp}.

COMPUTER PROJECT 2-3 | *Elementary Quantum Mechanics*

Once a numerical integration scheme that permits easy insertion of defined functions and convenient setting of the limits of integration has been set up and debugged, we may wish to use numerical integration for convenience even when it is not necessary to do so. For example, establishing that wave functions have been correctly normalized and distinguishing between normalized and nonnormalized wave functions is a common exercise in introductory quantum mechanics and can be mathematically difficult for all but the lowest orbitals. Simply typing FNA(X) into a program like INTSIM (disk) as the wave function in question, and integrating its square, approximates 1.0 for normalized wave functions and something else for nonnormalized functions. The limits of integration should be large relative to electronic excursions. Usually, the limits are fairly obvious, but if they are not, they can be systematically incremented until a self-consistent integral is found.

Evaluation of the integral $\int_{r_1}^{r_2} \psi^2(r)\,dr$, where $\psi(r)$ is a normalized radial wave function, yields the probability density for finding an electron in that orbital at a radial distance $r_1 < r < r_2$ from the nucleus. A common assigned problem in elementary quantum chemistry (McQuarrie, 1983; Hanna, 1981) is to determine the probability of finding an electron in the 1s orbital of hydrogen at a radial distance of 1 bohr radius (a_0) or less from the nucleus. This problem is usually solved by integration in closed form (*Ans.* $p = 0.323$) but the wave function can easily be introduced into an iterative procedure, such as a Simpson's rule integration program, that calculates the probability at any stipulated value of r:

$$P = \frac{4}{a_0^3} \int_0^r r^2 \exp\left(\frac{-2r}{a_0}\right) dr = 4 \int_0^x x^2 \exp(-2x)\,dx$$

where $x = r/a_0$.

Procedure Program ATOM is a modification of INTSIM that calculates the probability of finding an electron between any two radii (in units of r/a_0) in the $\psi_{1s} = e^{-r/a_0}$ orbital of hydrogen. Draw a cumulative probability curve (P vs. r/a_0) for finding an electron within any given radius. The curve resembles an *ogive* or S-shaped curve common in chemical applications but is flattened at the top owing to the nongaussian nature of the square of the 1s wave function. Does the curve pass through the point {0.323, 1}? Under what conditions does it attain its maximum value?

An extension of this project is to set up integration limits so that critical radii can be generated that contain the electron with a probability of $0.1, 0.2, \ldots, 0.9$. Knowing these radii, probability contour maps can be drawn (Gerhold, 1972). Draw the appropriate contour map for the hydrogen atom. Modify ATOM to compute P vs. r/a_0 for $\psi_{2s} = (2 - r/a_0)\exp(-r/2a_0)$. The normalizing factor $\frac{1}{8}$ replaces 4 in line 110. Based on this curve, does an excited 2s electron spend most of its time in the inner or outer shell of the hydrogenic 2s orbital?

COMPUTER PROJECT 2-4 | *Numerical Integration of Experimental Data Sets*

Thermodynamics and physical chemistry texts develop the equation

$$S_2 = S_1 + \int_{\ln T_1}^{\ln T_2} C_p\, d(\ln T) \tag{2-20}$$

where S is the entropy and C_p is the heat capacity at constant pressure. Equation (2-20) is the fundamental equation for determining the entropy change of a substance that is heated from T_1 to T_2 but does not suffer a phase change over that temperature interval. The alternative form

$$S_2 = S_1 + \int_{T_1}^{T_2} \frac{C_p}{T} \, dT \tag{2-21}$$

is also used. Armed with the third law of thermodynamics, heat capacities, and thermodynamic data that permit calculation of accurate entropies of intervening phase changes, these integrations permit one to determine absolute entropies. Several examples have been given (Norris, 1981) in which the entropy of a diatomic gas at 500 K is determined from a knowledge of its entropy at 298.15 K and numerical integration of accurate heat capacity data over that temperature range. Several other chemical applications of numerical integration are given, including determination of the equilibrium constant at an arbitrary temperature T_2 from the integrated van 't Hoff equation (Cox and Pilcher, 1970) and a knowledge of K_1 at T_1. Supporting algorithms, data tables, references, and commentaries on the calculations are given. In this project, the analytical form of the functional relationship is not used; integration is carried out directly from an experimental data set, necessitating the rather different approach of program ENTROPY (disk).

   Procedure Program ENTROPY has a self-contained data set. It computes the entropy of lead at 298 K by Simpson's rule. The data set consists of experimental values of C_p/T vs. T, where T is regarded as the independent variable. Because integration is from 0 to 298 K, S is an absolute entropy (third law of thermodynamics). Note that the data points are used in pairs; therefore, there must be an even number of them with a value of C_p/T and of T for each point.

   Run the program. Determine $S^{298}(Pb)$. Inspect the program listing to see how it works. What is the entropy of Pb at 100 K and at 200 K? Sketch the curve of C_p vs. T for lead. Sketch the curve of C_p/T vs. T.

   A lustrous metal has the following heat capacities as a function of temperature:

$$0.0, 5, 0.24, 10, 0.64, 15, 1.36, 20, 2.31, 25, 3.14,$$
$$30, 4.48, 50, 9.64, 70, 15.7, 100, 20.2, 150, 22.0,$$
$$200, 23.4, 250, 24.3, 298, 25.5$$

What is its entropy at 298 K? Scan a table of standard entropy values and decide what the metal might be.

COMPUTER PROJECT 2-5 | *Evaluation of Definite Integrals*

(a) The following definite integrals are widely used in the kinetic theory of gases. They can be evaluated by certain "trick" substitutions and coordinate changes, which is fine unless you do not happen to know the trick. Integrate these functions using the INTSIM program by making appropriate substitutions in the define function DEF FN statement. At the outset, let $a = 2$.

$$\int_0^\infty e^{-ax^2} dx$$

$$\int_\infty^\infty e^{-ax^2} dx$$

$$\int_0^\infty xe^{-ax^2} dx$$

$$\int_\infty^\infty xe^{-ax^2} dx$$

$$\int_0^\infty x^3 e^{-ax^2} dx$$

$$\int_\infty^\infty x^3 e^{-ax^2} dx$$

(b) A function for which $f(x) = -f(-x)$ is called an odd function. If $f(x) = f(-x)$, the function is even. Find some odd functions as the whole or parts of some of the preceding functions. Find some even functions. Find a general rule for the integrals of odd functions over a symmetrical interval. What about the integral of the product of an odd function and an even function over an interval that is symmetrical for both?

(c) Make systematic changes in a in the preceding integrals and try to induce an expression for the integrals that includes a as a parameter. For example,

$$\int_0^\infty x^2 e^{-ax^2} dx = \left(\frac{1}{4a}\right)\left(\frac{\pi}{a}\right)^{1/2}$$

(d) In Chapter 1, the "known integral"

$$\int_0^\infty \frac{x^3 \, dx}{e^x - 1} = \frac{\pi^4}{15}$$

was used. Verify this integral by numerical integration.

COMPUTER PROJECT 2-6| *The Dissociation Energy of* H_2^+

The hydrogen molecule ion H_2^+ is of enormous theoretical importance because the complete quantum-mechanical calculation of its dissociation energy can be carried out by both exact and approximate methods. This permits comparison of the exact quantum-mechanical solution with the solution obtained by various approximate techniques so that a judgment about the efficacy of the approximate methods can be made. In general, exact quantum-mechanical calculations cannot be carried out; hence, the importance of the one exact molecular solution we do have.

The experimental energy of dissociation of H_2^+ is equal to its bond energy. We wish to have a three-way comparison: (i) exact theoretical, (ii) approximate theoretical, and (iii) experimental. The exact solution is 2.791 eV (Hanna, 1981). Approximate calculations will be discussed in Chapter 12. Experimental determination of the dissociation energy of H_2^+ will be carried out in this project by analysis of spectroscopic data with the use of a Birge–Spooner plot. A simple mathematical manipulation of the vibration frequencies of H_2^+ yields the electronic dissociation energy we seek.

Method The vibrational spectrum of an ideal harmonic oscillator would consist of one line at frequency ν corresponding to $\Delta E = h\nu$, where ΔE is the same for any transition to an adjacent energy level (selection rule $\Delta n = \pm 1$) in an energy-level manifold that has all levels evenly spaced.

The H_2^+ is not a perfect harmonic oscillator. The vibrational levels for H_2^+ get closer together as the vibrational energy increases. Ultimately, a limit is reached at which the internuclear separation r increases by a large amount for an infinitesimal increase in potential energy V. At this limit, dissociation has occurred:

$$H_2^+ \rightarrow H + H^+$$

It is the energy required to bring about this reaction that we need in order to determine the bond energy of the hydrogen molecule ion. Spectroscopy permits us to calculate the energy difference between the lowest level and the next higher level from Planck's equation. The energy corresponding to this transition gives the spectral line of highest frequency because, of all energy spacings, the first is the largest. We can also measure the second energy increment, which corresponds to the spectral peak of next lower frequency, and so on. We have a series of energies, each smaller than the one before, corresponding to the gradual diminution of energy spacing in Fig. 2-6(b). The series approaches 0. The sum of all energies of transition is the energy of dissociation. The bond energy is the depth of the potential energy well in Fig. 2-6 and is closely related to the dissociation energy.

We can approximate the sum of energies as an integral:

$$\sum E = h \int \nu \, dn = \int E \, dn$$

FIGURE 2-6 The potential well and energy levels of (a) a perfect harmonic oscillator and (b) an anharmonic oscillator resembling H_2^+. D_0 is the dissociation energy and D_e is the bond energy. Frequencies of vibrational transitions in reciprocal centimeters: 2191, 2064, 1941, 1821, 1705, 1591, 1479, 1368, 1257, 1145, 1033, 918, 800, 677, 548, 411, 265, 117.

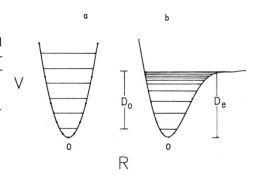

which is the area under the curve in Fig. 2-7, called a Birge–Spooner plot, in which n is the vibrational quantum number.

Procedure Run program DIS after having filled in the appropriate data block. This is not as easy as it sounds. The listing of the spectral lines given in the legend for Fig. 2-6 must be carefully thought out to provide the input data block that will give the right answer. Furthermore, the output of the program, in kilojoules per mole, is not exactly the answer we seek but will provide the right answer after one further mathematical step. If all calculations are carried out correctly, the result will be within 1% of the 2.791-eV literature value.

FIGURE 2-7 Birge–Spooner plot of the energy increment ΔE between vibrational energy levels vs. the vibrational quantum number n.

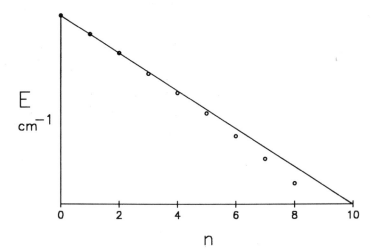

PROBLEMS | *Chapter 2*

1 The program in Exercise 2 gives final values for the integral under the normal curve that are obviously too large. The last entry is 0.5002 whereas, from the nature of the problem, we know that the integral cannot exceed 0.5000. Suggest a reason for this.

2 Show that the area under a parabolic arc that is concave upward is $\frac{1}{3}w(f(x_i) + 4f(x_{i+1}) + f(x_{i+2}))$.

3 Compute the probability of finding a randomly selected experimental measurement between the limits ± 0.5 standard deviations from the mean.

4 Given experimental measurements with $\mu = 123.4$ and $\sigma = 12.9$, draw the entire probability distribution curve for the population of all experimental measurements in the population studied.

5 Obtain Eq. (2-15) from Eq. (2-14).

6 Is the atomic wave function

$$\psi = \frac{1}{\pi^{1/2}} \exp(-r)$$

normalized to 1? Is

$$\psi = \frac{1}{(32\pi)^{1/2}} \left(\frac{Z}{a_0}\right)^{3/2} (2 - x)e^{-x/2}, \qquad x = \frac{Zr}{a_0}$$

normalized to 1? (a_0 = bohr radius = 0.529177 Å.)

7 What is the probability of finding an electron between 0.6 and 1.2 bohr radii of the nucleus. Assume the electron to be in the $1s$ orbital of hydrogen.

8 The $2s$ orbital of hydrogen can be written

$$\psi = (2 - r)e^{-r}$$

Plot this orbital with appropriate scale factors to determine the behavior of ψ in both rectangular and spherical polar coordinates.

9 Plot the square of ψ as a function of r in spherical space, i.e., simply square the preceding function with an appropriate scale factor as determined by trial and error. Comment about the relationship between your plot and the shell structure of the atom. (See also Chapter 11.)

10 Plot the probability of finding an electron in the $2s$ orbital of hydrogen at distance r from a hydrogen nucleus as a function of r. (See also Chapter 11.)

11 Draw the probability map for an electron in the $2s$ orbital of the hydrogen atom by finding the radii within which the probability of finding the electron is 0.2, 0.4, 0.6, and 0.8. How does this distribution differ from the $1s$ orbital? (See also Chapter 11.)

12 Determine the absolute entropy of nitrogen at 298 K from heat capacity data (Levine, 1983)

13 Equation (2-16) states that the expectation value for speeds is $\langle v \rangle = \int_a^b vG(v)\, dv$. This integral can be replaced by a sum for a discontinuous distribution. Let s be the score (2 through 12) achieved by throwing two dice. Calculate the expectation value of s from

$$\langle s \rangle = \sum sp(s)$$

where $p(s)$ is the probability of rolling score s. [*Hint*: The first term in the sum is $2(1/6)(1/6) = 2/36$.]

REFERENCES

Atkins, P. W., 1982. *Physical Chemistry*, 2nd ed. W. H. Freeman, New York, Chapter 25.

Balaam, L. N., 1972. *Fundamentals of Biometry*. Halsted (Wiley), New York, Chapters 5 and 6.

Cox, J. D. and Pilcher, G., 1970. *Thermochemistry of Organic and Organometallic Compounds*. Academic, London.

Gerhold, G., et al., 1972. *Am. J. Phys.*, **40**, 918.

Hanna, M. W., 1981. *Quantum Mechanics in Chemistry*, 3rd ed. Benjamin, Menlo Park, CA, Exercise 6-11, p. 133.

Kauzmann, W., 1966. *Kinetic Theory of Gases*. Benjamin, New York, Section 4-3. A detailed solution is to be found in the answer manual to Levine (1983), Problem 15-23.

Levine, I. N., 1983. *Physical Chemistry*. McGraw-Hill, New York, pp. 434–437.

McQuarrie, D. A., 1983. *Quantum Chemistry*. University Science Books, Mill Valley, CA, Example 6-10, p. 226.

Norris, A. C., 1981. *Computational Chemistry: An Introduction to Numerical Methods*. Wiley, New York.

Rice, J. R., 1983. *Numerical Methods, Software and Analysis*. McGraw-Hill, New York. An advanced treatment.

Rogers, D. W., 1983. *BASIC Microcomputing and Biostatistics*. Humana Press, Clifton, NJ

Rogers, D. W. and Gratzer, W., 1984. *Am. Lab.* 16(9), 78.

Scheid, F., 1968. *Numerical Analysis*. Schaum's Outline Series, McGraw-Hill, New York, Chapter 14.

Young, H. D., 1962. *Statistical Treatment of Experimental Data*. McGraw-Hill, New York.

3 | *Matrices*

A matrix is a rectangular array of elements, for example

$$\mathbf{A} = \begin{pmatrix} a_{11} & a_{12} \\ a_{21} & a_{22} \end{pmatrix} \tag{3-1}$$

Each element is designated using a double subscript; in general, an element is called a_{ij} where j is its horizontal position in the ith row of the matrix. A matrix with m rows and n elements in each row is an $m \times n$ matrix. A square matrix with n elements in each row is an $n \times n$ matrix.

Matrices obey an algebra of their own, which resembles the algebra of numbers in some respects and not in others. The elements of a matrix may be numbers, operators, or functions. We shall deal primarily with matrices of numbers in this chapter, but matrices of operators and functions will be important later.

Addition and subtraction of matrices is carried out by adding or subtracting corresponding elements. If we denote matrices by capital letters and elements by lowercase letters, and

$$\mathbf{C} = \mathbf{A} + \mathbf{B}$$

then each element in $\mathbf{C}$ is the sum of the corresponding elements in $\mathbf{A}$ and $\mathbf{B}$:

$$c_{ij} = a_{ij} + b_{ij}$$

It should be evident that there must be the same number of elements in two matrices to be added and that the elements must be arranged in the same way, so that there is a match of one element in matrix $\mathbf{A}$ with its corresponding element in matrix $\mathbf{B}$. Such matrices are said to be conformable to addition.

EXERCISE 3-1

Give an example of matrices that are conformable to addition and an example of matrices that are not.

Solution 3-1 The matrices

$$\begin{pmatrix} 2 & 7 \\ 1 & 1 \end{pmatrix} \quad \begin{pmatrix} -4 & -1 \\ 0 & -3 \end{pmatrix}$$

are conformable to addition. The matrices

$$\begin{pmatrix} 2 & 7 & 3 \\ 1 & 1 & 5 \end{pmatrix} \quad \begin{pmatrix} 4 & -1 \\ 0 & 3 \end{pmatrix}$$

are not conformable to addition. The matrices

$$\begin{pmatrix} 2 & 7 & 3 \\ 1 & 1 & 5 \end{pmatrix} \quad \begin{pmatrix} 4 & -1 \\ 0 & 3 \\ 9 & 1 \end{pmatrix}$$

are not conformable to addition.

Subtraction of matrices is analogous to addition. If

$$\mathbf{D} = \mathbf{A} - \mathbf{B}$$

then

$$d_{ij} = a_{ij} - b_{ij}$$

Matrices **A** and **B** must be conformable to subtraction.

The normal rules of association and commutation apply to matrices just as they apply to the algebra of numbers.

The zero matrix has 0 as all its elements; hence, addition to or subtraction from A leaves A unchanged:

$$\mathbf{A} + \mathbf{0} = \mathbf{A}$$

The zero matrix is sometimes called the *null matrix*.

MATRIX MULTIPLICATION

Multiplication of a matrix by a scalar x follows the rules one would extrapolate from the algebra of numbers: Each element of **A** is multiplied by the scalar. If

$$\mathbf{E} = x\mathbf{A}$$

then

$$e_{ij} = xa_{ij}$$

Multiplication of two matrices, however, is quite different from multiplication of two numbers. The first row of the premultiplying matrix is multiplied element by element into the first column of the postmultiplying matrix and the sum is the first element in the product matrix. This process is repeated with the first row of the premultiplying matrix and the second column of the postmultiplying matrix to obtain the second element in the product matrix and so on, until all of the elements of the product matrix have been filled in. If

$$\mathbf{F} = \mathbf{AB}$$

then

$$f_{ij} = \sum_{k=1}^{n} a_{ik} b_{kj} \tag{3-2}$$

To be conformable to multiplication, the horizontal dimension of **A** must be the same as the vertical dimension of **B**, that is, $n_A = m_B$. Also, m_A must be the same as n_B. Square matrices of the same size are, of course, always conformable to multiplication. This unusual definition of multiplication with its rules for dimensions will become clear with repeated use. The matrices we shall be interested in will usually be square. Assume that the matrices we will discuss are square unless stipulated otherwise. The rules for rectangular matrices and column and row matrices will be developed as needed.

EXERCISE 3-2

Find the product **AB** where

$$A = \begin{pmatrix} 1 & 2 \\ 3 & 4 \end{pmatrix} \qquad B = \begin{pmatrix} 5 & 6 \\ 7 & 8 \end{pmatrix}$$

Solution 3-2

$$C = AB = \begin{pmatrix} 19 & 22 \\ 43 & 50 \end{pmatrix}$$

Matrix multiplication is not commutative, that is,

$$AB \neq BA_{\text{general case}} \tag{3-3}$$

DIVISION OF MATRICES

Division of matrices is not defined as such, but the equivalent operation of multiplication by an inverse matrix (if it exists) is defined. If a matrix **A** is multiplied by its own inverse, A^{-1}, the *unit matrix* is obtained. The unit matrix has $1s$ on its principal diagonal (the longest diagonal from upper left to lower right) and 0s elsewhere, e.g.,

$$I = \begin{pmatrix} 1 & 0 & 0 \\ 0 & 1 & 0 \\ 0 & 0 & 1 \end{pmatrix}$$

It plays the role in matrix algebra that 1 plays in ordinary algebra. Multiplication of a matrix by the unit matrix leaves it unchanged:

$$AI = A$$

Inverse matrices commute:

$$AA^{-1} = A^{-1}A = I \tag{3-4}$$

Among the ordinary numbers, only 0 has no inverse. Many matrices have no inverse. The question of whether a matrix **A** has or does not have a defined inverse is closely related to the question of whether a set of simultaneous equations has or does not have a unique set of solutions. We shall consider this question more fully later, but for now, recall that if one equation in a pair of simultaneous equations is a multiple of the other,

$$x + 2y = 4$$

$$2x + 4y = 8$$

no unique solution exists. Similarly for matrices, if one row (or column) of elements is a multiple of any other in the matrix, e.g.,

$$\mathbf{A} = \begin{pmatrix} 1 & 2 \\ 2 & 4 \end{pmatrix}$$

no inverse exists.

EXERCISE 3-3

Obtain the product matrix **AB** where

$$\mathbf{A} = \begin{pmatrix} 1 & 2 & 3 \\ 4 & 5 & 6 \\ 7 & 8 & 9 \end{pmatrix}$$

and

$$\mathbf{B} = \begin{pmatrix} 1 & 0 & 1 \\ 2 & 1 & 2 \\ 4 & 1 & 3 \end{pmatrix}$$

The operation requires 27 individual multiplications and 9 additions.

EXERCISE 3-4

Solve for **AB** above using program MMULT. Solve for **BA**. Do **AB** and **BA** commute?

Powers and Roots of Matrices

If two square matrices of the same size can be multiplied, then a square matrix can be multiplied into itself to obtain $\mathbf{A}^2$, $\mathbf{A}^3$, or, in general, $\mathbf{A}^n$. **A** is the square root of $\mathbf{A}^2$ and **A** is the nth root of $\mathbf{A}^n$. A number has only two square roots, but a matrix has infinitely many square roots. This will be demonstrated in the problems at the end of this chapter.

Matrix Polynomials

Polynomial means many terms. Now that we are able to multiply a matrix by a scalar and find powers of matrices, we can form matrix polynomials, e.g.,

$$\mathbf{A}^2 + 4\mathbf{A} + 5\mathbf{I} = 0 \tag{3-5}$$

There are infinitely many matrices that satisfy this polynomial, hence the polynomial has infinitely many roots.

EXERCISE 3-5

Show that the matrix

$$\mathbf{A} = \begin{pmatrix} 2 & 3 \\ 3 & 2 \end{pmatrix}$$

satisfies the polynomial

$$\mathbf{A}^2 - 4\mathbf{A} - 5\mathbf{I} = 0$$

EXERCISE 3-6

Show that the matrix

$$\mathbf{A} = \begin{pmatrix} 2 & 3x \\ 3/x & 2 \end{pmatrix}$$

satisfies the equation

$$\mathbf{A}^2 - 4\mathbf{A} - 5\mathbf{I} = 0$$

for any value of x. How many roots does this equation have?

EXERCISE 3-7

What are the roots of the ordinary polynomial

$$a^2 - 4a - 5 = 0$$

EXERCISE 3-8

Notice that the polynomial in $\mathbf{A}$ can be factored to give

$$(\mathbf{A} - 5\mathbf{I}) \quad \text{and} \quad (\mathbf{A} + \mathbf{I})$$

Perform the subtraction and addition as prescribed and multiply the resultant matrices to show that the null matrix is obtained thereby.

The general form for a matrix polynomial satisfied by $\mathbf{A}$ is

$$c_m\mathbf{A}^m + c_{m-1}\mathbf{A}^{m-1} + \cdots + c_0\mathbf{I} = 0 \qquad (3\text{-}6)$$

The Least Equation

The least equation is the polynomial satisfied by $\mathbf{A}$ that has the smallest possible degree. There is only one least equation:

$$\mathbf{A}^k + c_{k-1}\mathbf{A}^{k-1} + \cdots + c_0\mathbf{I} = 0 \qquad (3\text{-}7)$$

The degree of the least equation k is called the rank of the matrix $\mathbf{A}$. If $k = n$, the size of a square matrix, the inverse $\mathbf{A}^{-1}$ exists. The degree k is never greater than n for the least equation (although there are other equations that $\mathbf{A}$ satisfies for which $k > n$). If the matrix is not square or $k < n$, then $\mathbf{A}$ has no inverse.

One method of finding the least equation is illustrated here for the simplest second degree case. Find a number r such that

$$\mathbf{A}^2 - r\mathbf{I}$$

is a matrix that has 0 as the lead element (the element in the $1, 1$ position). Now, find a number s such that

$$\mathbf{A} - s\mathbf{I}$$

has 0 as the lead element. Find a number t such that

$$(\mathbf{A}^2 - r\mathbf{I}) - t(\mathbf{A} - s\mathbf{I}) = 0$$

This leads to the least equation

$$\mathbf{A}^2 - t\mathbf{A} + (ts - r)\mathbf{I} = 0$$

where the coefficients $c_0 = ts - r$ and $c_1 = -t$ in Eq. (3-7).

If the coefficient $c_0 = 0$, the matrix $\mathbf{A}$ is *singular* and has no inverse. The method can be extended to higher degrees but becomes tedious.

EXERCISE 3-9

Use the preceding method to find the least equation of the matrix

$$\mathbf{A} = \begin{pmatrix} 2 & 1 \\ 1 & 3 \end{pmatrix}$$

Does $\mathbf{A}$ have an inverse?

Solution 3-9

$$\mathbf{A}^2 - 5\mathbf{A} + 5\mathbf{I} = 0$$
$$c_0 \neq 0; \qquad \mathbf{A}^{-1} \text{ exists}$$

Verify this solution by calculating and substituting $\mathbf{A}^2$ and $5\mathbf{A}$ to prove the equality.

The *adjoint method* of finding inverse matrices relies on the equality

$$\mathbf{A}^{-1} = -1/c_0\left(\mathbf{A}^{k-1} + \cdots + c_1\mathbf{I}\right) \tag{3-8}$$

EXERCISE 3-10

Determine $\mathbf{A}^{-1}$ by the adjoint method and calculate the products $\mathbf{A}^{-1}\mathbf{A}$ and $\mathbf{A}\mathbf{A}^{-1}$. Are they both equal to $\mathbf{I}$?

The adjoint method results in some work reduction in finding the inverse of a small matrix by hand. In this book, we shall prefer numerical computational methods. Computing inverse matrices will be a major theme of this book.

Importance of Rank

The degree of the least polynomial of a square matrix $\mathbf{A}$, hence its rank, is the number of independent rows in $\mathbf{A}$. An independent row of $\mathbf{A}$ is a row that cannot be obtained from any other row in $\mathbf{A}$ by multiplication by a number. If matrix $\mathbf{A}$ has as its elements the coefficients of a set of simultaneous nonhomogeneous equations (see Chapter 4), the rank k is the number of independent equations. If $k = n$, there are the same number of independent equations as unknowns; $\mathbf{A}$ has an inverse and a unique solution set exists. If $k < n$, the number of independent equations is less than the number of unknowns; $\mathbf{A}$ does not have an inverse and no unique solution set exists. The matrix $\mathbf{A}$ is square; hence, $k > n$ is not possible.

Importance of the Least Equation

A number s for which

$$\mathbf{A} - s\mathbf{I}$$

has no reciprocal is called an *eigenvalue* of $\mathbf{A}$. The equation

$$\mathbf{A}\mathbf{V} = s\mathbf{V}$$

where $\mathbf{V}$ is a vector (or vector function) is called the eigenvalue equation. If

$$\mathbf{A}^k + c_{k-1}\mathbf{A}^{k-1} + \cdots + c_0\mathbf{I} = 0$$

is the least equation satisfied by $\mathbf{A}$, then s is an eigenvalue only if

$$s^k + c_{k-1}s^{k-1} + \cdots + c_0 = 0$$

This is one way of finding eigenvalues. All atomic and molecular energy levels are eigenvalues of a special eigenvalue equation called the *Schroedinger equation*.

EXERCISE 3-11

Perform the matrix subtraction

$$A - EI$$

where

$$A = \begin{pmatrix} \alpha & \beta \\ \beta & \alpha \end{pmatrix}$$

What is the condition on the resulting matrix that must be met if E is to be an eigenvalue of A?

Solution 3-11 The matrix

$$A = \begin{pmatrix} \alpha - E & \beta \\ \beta & \alpha - E \end{pmatrix}$$

must have no inverse.

Special Matrices

The transpose A^t of a matrix is obtained by reflecting the matrix through its principal diagonal:

$$a_{ij}^t = a_{ji}$$

Properties of the transpose include

$$(A + B)^t = A^t + B^t$$

and

$$(AB)^t = B^t A^t$$

(note the order of A and B).

EXERCISE 3-12

Demonstrate that these properties hold for arbitrarily selected matrices A and B.

A *symmetric* matrix equals its own transpose.

Historical Note: It is interesting to note (Pauling and Wilson, 1935) that the very first systematic approach to what we now call quantum mechanics was made by Heisenberg, who began to develop his own algebra to describe the frequencies and intensities of spectral transitions. It was soon seen by Born and Jordan that the new algebra is really matrix algebra. Heisenberg's functions were later called wave functions by Schroedinger in his independent but equivalent method. Schroedinger's method is called wave mechanics and is the method most familiar to chemists. Heisenberg's method is called matrix mechanics.

EXERCISE 3-13

Give three examples of symmetric matrices.

The *trace* of a matrix is the sum of the elements on its principal diagonal

$$\text{tr}(\mathbf{A}) = \sum a_{ii}$$

EXERCISE 3-14

What is the trace of a unit matrix of size n?

A *diagonal* matrix has nonzero elements only on the principal diagonal and zeros elsewhere. The unit matrix is a diagonal matrix. *Bidiagonal* and *tridiagonal* matrices with two or three diagonals nonzero are not uncommon in computational chemistry, nor are large matrices with small matrices symmetrically lined up along the principal diagonal.

Triangular matrices have nonzero elements only on and above the principal diagonal (upper triangular) or on and below the principal diagonal (lower triangular). Some of the more important computational methods are devoted to transforming a general matrix into its equivalent diagonal or triangular form.

MACHINE CONSIDERATIONS

Decimal numbers must be transformed into binary numbers before the microprocessor can work on them. A binary number is an ordered sequence of 0 or 1 called *bits*. The entire collection of bits is called a *word*. A word might look like this: 10101101. Rounding error is encountered whenever the word size required for a calculation is larger than the number of bits that can be handled by the microprocessor. In this event, extra bits may simply be dropped, leaving the truncated number as the final or intermediate answer to the problem. The truncated binary number is then translated back to decimal for numeric output. Some systems provide an algorithm for rounding decimal numbers up or down, according to the usual rules of number handling. In many simple calculations, this procedure is adequate because numbers are returned with six digits or so of accuracy, which is more accurate than most experimental measurements.

One cannot be sanguine about rounding error, however, because there are some very accurate experimental measurements that would suffer if calculated to six digits of accuracy. Many calculated results are the difference between two large computed values and are sensitive to small errors in either or both. Also, the error of many repetitive computations accumulates on each iteration. Often divergence in a cumulative calculation gets worse as the number of iterations increases, though oscillation about the right answer is also common. The reader should use common sense in interpreting a computed result. A calculation that produces 1.00000, 7.81942×10^{-7}, -0.99999 may correctly be said to have returned the answer 1, 0, -1.

Programs MMULT and MINV (disk) have written data sets. To solve new problems, simply overwrite the data statements, e.g., 30 DATA x, x, x (new numbers) and save the modified program to a unique file name if desired. Matrices of dimension other than 3×3 can be multiplied using an altered value of N, e.g., 40 $N = 4$. The DIM statement clears memory for a *maximum* dimension of 30. It need not be changed for smaller matrices. One would normally use a mainframe computer for matrices approaching 30×30.

COMPUTER PROJECT 3-1 | *Matrix Manipulations*

The matrix

$$A = \begin{pmatrix} 1 & 2 \\ 3 & 4 \end{pmatrix}$$

satisfies the equation

$$A^2 - 5A - 2I = 0$$

whence

$$A^2 = 5A + 2I$$
$$A^4 = (5A + 2I)^2$$

and

$$A^8 = (5A + 2I)^4$$

Matrix algebra yields the same simplification for A^4 that one would get from ordinary algebra

$$A^4 = (5A + 2I)^2$$
$$= 25A^2 + 20A + 4I$$
$$= 25(5A + 2I) + 20A + 4I$$
$$= 145A + 54I$$

(a) Show that

$$A^8 = 120785A + 44966I$$

This is a convenient way of obtaining even powers of a matrix. Find A^8 using this equation. This accurate method of obtaining A^8 should yield the true value; we shall call the matrix obtained this way the "true" $\mathbf{A}^8$ and designate it $\mathbf{A}^8_t$.

(b) Find $\mathbf{A}^6$. Use the rule of exponents from ordinary algebra that $a^x(a^y) = a^{x+y}$. We will cross-check the answer later.

(c) Find $\mathbf{A}^{-1}$ and multiply

$$\mathbf{A}^{-1}(\mathbf{A}^8) = \mathbf{A}^7$$

Do $\mathbf{A}^{-1}$ and $\mathbf{A}^8$ commute? This is a convenient way to find odd powers of a matrix.

(d) Compute the matrix $\mathbf{A}^8$ by the "brute force and ignorance" method of using a computer multiplying routine. (Modify your multiplication program MMULT so that it loops eight times.) We shall call the matrix obtained this way $\mathbf{A}^8_{bfi}$

$$\mathbf{A}\,\mathbf{A}\,\mathbf{A}\,\mathbf{A}\,\mathbf{A}\,\mathbf{A}\,\mathbf{A}\,\mathbf{A} = \mathbf{A}^8_{bfi}$$

Find the difference

$$\mathbf{A}^8_t - \mathbf{A}^8_{bfi} = \mathbf{E}$$

This difference is the error matrix $\mathbf{E}$, which should be zero. It may not be, however, due to computer rounding error.

(e) Suppose the original matrix $\mathbf{A}$ represents a set of experimental measurements in which an error of 0.1% was made in measuring the first element a_{11}. Now

$$\mathbf{A}_{exper} = \begin{pmatrix} 1.001 & 2 \\ 3 & 4 \end{pmatrix}$$

owing to experimental uncertainty. The new error matrix of $\mathbf{A}$ is defined as

$$\mathbf{E} = \mathbf{A}_{exper} - \mathbf{A}_t$$

The error matrices of the eighth powers of $\mathbf{A}$ are

$$\mathbf{E}' = \mathbf{A}^8_{exper} - \mathbf{A}^8_t$$

$$\mathbf{E}'_{bfi} = \left(\mathbf{A}^8_{exper}\right)_{bfi} - \mathbf{A}^8_t$$

Calculate the error matrix $\mathbf{E}'$ and $\mathbf{E}'_{bfi}$ using the polynomial and brute force methods, where we already know $\mathbf{A}_t$. Compare and comment upon $\mathbf{E}$ and $\mathbf{E}_{bfi}$.

Nothing in this section should be construed to mean that brute force and ignorance methods should never be used. Indeed, the computer's great strength is its ability to do simple things fast. One must, however, be much more aware of cumulative error in a computer calculation that involves many steps than in a hand calculation that involves only a few.

PROBLEMS | *Chapter* 3

1 Find the product **AB**:

$$\mathbf{A} = \begin{pmatrix} 3 & 0 & 3 \\ 4 & -1 & -1 \\ 1 & 2 & 5 \end{pmatrix} \qquad \mathbf{B} = \begin{pmatrix} 1 & 1 & 1 \\ -2 & 1 & 6 \\ 3 & 4 & 5 \end{pmatrix}$$

2 Invert **A** in Problem 1. Systematic methods exist for inverting matrices and will be discussed in the next topic. For now, use the matrix programs MMULT and MINV to solve this problem set.

3 Find $\mathbf{AA}^{-1}$. Is it true that $\mathbf{AA}^{-1} = \mathbf{I}$? Do you encounter rounding error? (A statement that is true within 0.1% rounding error or less should be considered true for these problems.)

4 Find $\mathbf{A}^{-1}\mathbf{A}$. Is it true that $\mathbf{A}^{-1}\mathbf{A} = \mathbf{I}$?

5 Can rounding error be reduced by including more digits in the input DATA statement (going from, say, 0.166 to 0.16666)? Is there any limit to this?

6 Invert **B** in Problem 1.

7 Find the product $\mathbf{A}^{-1}\mathbf{B}^{-1}$ using the matrices in Problem 1.

8 Find the inverse of the product $\mathbf{C} = \mathbf{AB}$. Is it true that $(\mathbf{AB})^{-1} = \mathbf{A}^{-1}\mathbf{B}^{-1}$?

9 Is it true that $(\mathbf{BA})^{-1} = \mathbf{B}^{-1}\mathbf{A}^{-1}$? Is it true that $(\mathbf{AB})^{-1} = (\mathbf{BA})^{-1}$? Is it true that $(\mathbf{AB})^{-1} = \mathbf{B}^{-1}\mathbf{A}^{-1}$? Is it true that $(\mathbf{BA})^{-1} = \mathbf{A}^{-1}\mathbf{B}^{-1}$?

10 Transpose **A** and **B** in Problem 1.

11 Transpose the product **AB** to find $(\mathbf{AB})^t$.

12 Find the product $\mathbf{A}^t\mathbf{B}^t$. Compare it with the transpose of the product from Problem 11.

13 What is the transpose of the upper triangular matrix

$$\begin{pmatrix} a & b \\ & c \end{pmatrix}$$

REFERENCES

Barrante, J. R., 1974. *Applied Mathematics for Physical Chemistry.* Prentice-Hall, Englewood Cliffs, N.J.

Pauling, L. and Wilson, E. B., 1935. *Introduction to Quantum Mechanics.* McGraw-Hill, New York. Reprinted (1963), Dover, New York, Chapter XV.

Schwartz, J. T., 1961. *Introduction to Matrices and Vectors.* Dover, New York.

4

Linear Nonhomogeneous Simultaneous Equations

This chapter describes four common methods for solving sets of linear nonhomogeneous simultaneous equations (Scheid, 1968; Acton, 1970) and points out some of the similarities and differences that are important when choosing an algorithm for use with a personal microcomputer. Many coded programs have been published in FORTRAN (e.g., Carnahan, Luther, and Wilkes 1969; Isenhour and Jurs, 1979) for each of the algorithms discussed here. Most are short and easily translated into other computer languages.

Of interest to us here is the simple and restricted problem of n linear independent nonhomogeneous equations in n real unknowns. It is a problem very often encountered in an experimental context. The problem of more than n equations in n real unknowns is called the multivariate problem and will be treated in Chapter 6. The problem of linearly dependent *homogeneous* equations occurs frequently in some branches of quantum mechanics and will be treated in detail in Chapter 7.

Nomenclature

Taking the two-variable problem for notational simplicity but implying that it may be extended to the n-variable case, the equation set we describe may be written

$$a_{11}x_1 + a_{12}x_2 = b_1$$
$$a_{21}x_1 + a_{22}x_2 = b_2$$

(4-1)

Linear independence implies that no equation in the set can be obtained by multiplying any other equation in the set by a constant. The $n \times n$ matrix populated by n^2 elements a_{ij}

$$\begin{pmatrix} a_{11} & a_{12} \\ a_{21} & a_{22} \end{pmatrix}$$

is called the *matrix of the coefficients*. The principal diagonal runs from top left to bottom right.

An ordered set of numbers is a vector; hence, $X = \{x_1, x_2\}$ is called the *solution vector* or *solution set* and the ordered set of constants $\{b_1, b_2\}$

is called the *constant vector*. I like the term *nonhomogeneous vector*, because existence of any nonzero element b_i causes the equation set to be nonhomogeneous. A convenient term is the *B vector*. To be conformable for multiplication by a matrix, the dimension of a row vector must be the same as the column dimension of the matrix. A column vector must have the same dimension as the row dimension of the matrix. A few pencil-and-paper multiplications will clarify this rule. A vector is a matrix with 1 as one of its dimensions.

EXERCISE 4-1

Show that a vector in a plane can be unambiguously represented by an ordered number pair and, hence, that any ordered number pair can be regarded as a vector.

Solution 4-1 Consider a vector as an arrow in two-dimensional space (see Fig. 4-1). Now superimpose x-y coordinates on the two space, arbitrarily placing the origin on the tail of the arrow.

The vector happens to fall in the second quadrant as drawn. The number pair giving the point that coincides with the tip of the arrow gives its magnitude and direction relative to the coordinate system chosen. Magnitude and direction are all that one can know about a vector; hence, it is completely defined by the number pair $(5, -1)$.

In general, a vector in an n space can be represented by an n tuple of numbers, for example, a vector in three space can be represented as a number triplet.

The *determinant* having the same form as matrix **A**,

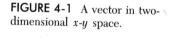

$$\mathbf{A} = \begin{vmatrix} a_{11} & a_{12} \\ a_{21} & a_{22} \end{vmatrix}$$

is not the same as **A** because a matrix is an operator and a determinant is a

FIGURE 4-1 A vector in two-dimensional x-y space.

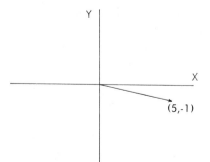

scalar; a matrix is irreducible but a determinant can, if it satisfies some conditions, be written as a single number. An elementary method of reducing a determinant to a single number is the *method of cofactors* (Anderson, 1966), but we shall not use it here because a better method exists for computers.

Designating the two vectors and one matrix just defined by capital letters, the set of equations [Eqs. (4-1)] is

$$\mathbf{AX} = \mathbf{B} \qquad (4\text{-}2)$$

Equation (4-2) is a matrix equation because vectors (**X** and **B**) are properly regarded as one-column matrices.

Multiplication by the rules of matrix algebra produces Eqs. (4-1) from Eq. (4-2) in ordinary algebraic form, demonstrating their equivalence. Equation (4-2) is an economical way of expressing Eqs. (4-1), especially where n is large, but it is more than that; systematic methods of solving Eqs. (4-1) really depend on the properties of the coefficient matrix and on what we can do with it. For example, if Eqs. (4-1) are linearly dependent, **A** is *singular*, which means, among other things, that its determinant is zero and it has no inverse $\mathbf{A}^{-1}$. In practical terms, this means that no unique solution set exists for Eqs. (4-1). We already knew that, but less obvious operations on Eqs. (4-1), such as triangularization and diagonalization, can be more easily visualized and programmed in terms of operations on the coefficient matrix **A** than in terms of the entire set.

ALGORITHMS

Although there are many methods of simultaneous equation solving by computer in the literature, they can be separated into two classes: elimination and iterative substitution. Elimination methods are closed methods: In principle, they are capable of infinite accuracy. Iterative methods converge on the solution set and so, strictly speaking, are never more than approximations to it. In practice, the distinction is not as great as it might seem, because iterative approximations can be made highly self-consistent, that is, identical from one iteration to another within the accuracy of a microcomputer, and closed elimination methods suffer the same machine word-size limitations that prevent infinite accuracy in any fairly involved microcomputer procedure.

ELIMINATION METHODS

Gaussian Elimination

In the most elementary use of *gaussian elimination*, the first of a pair of simultaneous equations is multiplied by a constant so as to make one of its coefficients equal to the corresponding coefficient in the second equation. Subtraction eliminates one term in the second equation, permitting solution of the equation pair.

Solving several equations by the method of gaussian elimination, one might divide the first equation by a_{11}, thus obtaining 1 in the a_{11} position. Multiplying a_{21} into the first equation makes $a_{11} = a_{21}$. Now subtracting the first equation from the second, a 0 is produced in the a_{21} position. The same thing can be done to produce a 0 in the a_{31} position and so on, until the first column of the coefficient matrix is filled with 0s except for the a_{11} position.

Attacking the a_{22} position in the same way, but leaving the first horizontal row of the coefficient matrix alone, yields a matrix with 0s in the first two columns except for the triangle

$$\begin{matrix} a_{11} & a_{12} \\ & a_{22} \end{matrix}$$

This is continued $n - 1$ times until the entire coefficient matrix has been converted to an *upper triangular matrix*, that is, a matrix with only 0s below the principal diagonal. The B vector is operated on in the same way as the coefficient matrix. The last equation, $a_{nn}x_n = b_n$, can be solved for x_n, which is back-substituted into the equation above it to obtain x_{n-1} and so on until the entire solution set has been generated.

EXERCISE 4-2

In Exercise 3-9, we obtained the least equation of the matrix

$$\begin{pmatrix} 2 & 1 \\ 1 & 3 \end{pmatrix}$$

and, by the adjoint method, we obtained the inverse $\mathbf{A}^{-1}$. Solve the simultaneous equation set by gaussian elimination

$$2x + y = 4$$
$$x + 3y = 7$$

Note that the matrix from Exercise 3-9 is the matrix of coefficients in this simultaneous equation set. Note also the similarity in method between finding the least equation and gaussian elimination.

Solution 4-2 The triangular matrix is

$$\begin{pmatrix} 1 & 0.5 \\ 0 & 2.5 \end{pmatrix}$$

and the nonhomogeneous vector has been transformed to $\{2, 5\}$. The bottom equation is $2.5y = 5$, i.e., $y = 2$. Back-substitution into the top equation yields $x = 1$. The solution set, as one could have seen by inspection, is $\{1, 2\}$.

In the computer algorithm, division by the diagonal element, multiplication and subtraction are usually carried out at the same time on each target element in the coefficient matrix, leading to some term like $a_{jk} - a_{ik}(a_{ji}/a_{ii})$. Next, the same three combined operations are carried out on the elements of the B vector. The arithmetic statements are simple, as is the procedure for back-substitution. The trick in writing a successful gaussian elimination program is in constructing a looping structure and keeping the variable indices straight so that the right operations are being carried out on the right elements in the right sequence.

EXERCISE 4-3

Test program GAUSEL by using it to solve the equation set in Exercise 4-2. The existing data set will have to be replaced by typing over the DATA statements with the new set.

Gauss–Jordan Elimination

Because it is possible to eliminate terms to obtain an upper triangular coefficient matrix, it is also possible to continue the process and eliminate the elements above the principal diagonal, leaving only those $a_{ii} \neq 0$ in the coefficient matrix. This extension of the gaussian elimination method is called the *Gauss–Jordan* method. It is easy and advantageous to switch the largest element in each row of the coefficient matrix into the pivotal position a_{ii}, and most Gauss–Jordan programs do this. Once the coefficient matrix has been *diagonalized* so that the a_{ii} are the only nonzero elements, and the same operations have been carried out on the B vector, the original system of n equations in n unknowns has been reduced to n equations, each in *one* unknown. The solution set follows routinely.

EXERCISE 4-4

Extend the matrix triangularization procedure in Exercise 4-2 by the Gauss–Jordan procedure to obtain the fully diagonalized matrix and the solution set. Do this calculation by hand.

Solution 4-4 The extended procedure yields

$$\begin{pmatrix} 1 & 0 \\ 0 & 2.5/5 \end{pmatrix}$$

as the diagonal matrix with the nonhomogeneous vector now given as $\{1, 1\}$. The solution set $\{1, 2\}$ follows from either the triangular or diagonal form.

Cramer's Rule

By Cramer's rule, each solution of Eqs. (4-1) is given as the ratio of determinants

$$x_i = D_i/D \qquad (4\text{-}3)$$

where D is the determinant of the coefficients a_{ij} and D_i is a similar determinant in which the ith column has been replaced by the elements of the B vector. The method is open-ended, that is, it can be applied to any number of equations containing the same number of unknowns, resulting in $n + 1$ determinants of dimension $n \times n$.

EXERCISE 4-5

Solve the equation set of Exercise 4-1 using Cramer's rule.

Although apparently quite different from the Gauss and Gauss–Jordan methods, it turns out that the most efficient method of reducing large determinants is mathematically equivalent to gaussian elimination. As far as the computer programming is concerned, the method of Cramer's rule is only a variant on the Gauss elimination method. It is slower because it requires evaluation of several determinants rather than triangularization or diagonalization of one matrix; hence, it is not favored, except where the determinants are needed for something else.

Determinants have one property that is very important in what will follow. If a row is exchanged with another row or a column is exchanged with another column, the determinant changes sign.

EXERCISE 4-6

Verify the preceding statement for the determinant

$$\begin{vmatrix} 1 & 2 \\ 3 & 4 \end{vmatrix}$$

Solution 4-6

$$\begin{vmatrix} 1 & 2 \\ 3 & 4 \end{vmatrix} = 4 - 6 = -2$$

$$\begin{vmatrix} 2 & 1 \\ 4 & 3 \end{vmatrix} = 6 - 4 = 2$$

ITERATIVE METHODS

Gauss–Seidel Iteration

The very first algorithm for Eqs. (4-1) that we learn in elementary algebra is usually a substitution method in which one variable is expressed explicitly as a function of the other and is substituted for in the remaining equation so that it is eliminated from the set. Back-substitution leads to the other member of the two-element solution set.

Substitution is not widely used with computers, but a variation called the *Gauss–Seidel* iterative method uses substitution in a way that is well-suited to machine computation and has the advantage of conceptual simplicity and ease of coding. In effect, one guesses a solution for x_1, substitutes this guess into the first equation, and solves for x_2. The solution is, of course, wrong, but it is substituted into the second equation to give a solution for x_1. That solution is also wrong, but, under some circumstances, it is less wrong than the original guess. The new approximation to x_1 is substituted to obtain a new x_2 and so on, in an iterative loop, until self-consistency is obtained, i.e., until the new approximation agrees with the old approximation to within some very small limit.

The drawback of the Gauss–Seidel method is that the iterative series does not always converge. Nonconvergence can be spotted by printing the approximate solution on each iteration. A favorable condition for convergence is dominance of the principal diagonal, that is, the term a_{ii} should be larger than the other terms in the ith row. A more detailed discussion of convergence is given in advanced texts (Rice, 1983; Norris, 1981).

MATRIX INVERSION
AND DIAGONALIZATION

Looking at matrix Eq. (4-2), one would be tempted to divide both sides by matrix **A** to obtain the solution set $\mathbf{X} = \mathbf{B}/\mathbf{A}$. Unfortunately, division by a matrix is not defined, but for some matrices, including nonsingular coefficient matrices, the inverse of **A** is defined.

The *unity matrix* **I** with $a_{ii} = 1$ and $a_{ij} = 0$ for $i \neq j$ plays the same role in matrix algebra that the number 1 plays in ordinary algebra. In ordinary algebra, we can perform an operation on any number, say 5, to reduce it to 1 (divide by 5). If we do the same operation on 1, we obtain the inverse of the original number, $1/5$. Analogously, in matrix algebra, if we carry out a *series* of operations on **A** to reduce it to the unity matrix and carry out the same series of operations on the unity matrix itself, we obtain the inverse of the original matrix $\mathbf{A}^{-1}$.

One series of mathematical operations that may be carried out on the coefficient matrix to diagonalize it is the Gauss–Jordan procedure. If each row is then divided by a_{ii}, the unity matrix is obtained. Generally, **A** and the unity matrix are subjected to identical *row operations* such that as **A** is reduced to **I**, **I** is simultaneously converted to $\mathbf{A}^{-1}$. The computer program that is written to do this is essentially a Gauss–Jordan program as far as coding and machine considerations are concerned (Isenhour and Jurs, 1979). Conversely, both reduction of **A** to **I** and conversion of **I** to $\mathbf{A}^{-1}$ may be done by the Gauss–Seidel iterative method (Noggle, 1985).

The attractive feature in matrix inversion is seen by premultiplying both sides of **AX** = **B** by $\mathbf{A}^{-1}$:

$$\mathbf{A}^{-1}\mathbf{A}\mathbf{X} = \mathbf{A}^{-1}\mathbf{B}$$

Because multiplying a matrix by its inverse gives unity, $\mathbf{A}^{-1}\mathbf{A} = \mathbf{I}$, and multiplying a matrix by the unity matrix leaves it unchanged,

$$\mathbf{I}\mathbf{X} = \mathbf{X} = \mathbf{A}^{-1}\mathbf{B}$$

This means that once $\mathbf{A}^{-1}$ is known, it can be multiplied into any number of *B* vectors to generate a solution set **X** for each. It is easier and faster to multiply a matrix into a set of vectors than it is to solve a set of simultaneous equations over and over for the same coefficient matrix but different *B* vectors.

EXERCISE 4-7

After reduction to the diagonal form by the Gauss–Jordan method, the equation set

$$2x + y = 4$$
$$x + 3y = 7$$

yielded

$$\begin{pmatrix} 1 & 0 \\ 0 & 0.5 \end{pmatrix} \begin{pmatrix} x \\ y \end{pmatrix} = \begin{pmatrix} 1 \\ 1 \end{pmatrix}$$

or

$$x + 0 = 1$$
$$0 + 0.5y = 1 \qquad \{x, y\} = \{1, 2\}$$

Now, place the original coefficient matrix next to the unit matrix

$$\begin{pmatrix} 2 & 1 \\ 1 & 3 \end{pmatrix} \begin{pmatrix} 1 & 0 \\ 0 & 1 \end{pmatrix}$$

Perform the necessary operations to diagonalize the coefficient matrix and at the same time perform the same operations on the adjacent unit matrix. Upon completion of this stepwise procedure, multiply the bottom row by 2 so that you have the unit matrix on the left and some matrix other than the unit matrix on the right:

$$\begin{pmatrix} 1 & 0 \\ 0 & 1 \end{pmatrix} \begin{pmatrix} z_{11} & z_{12} \\ z_{21} & z_{22} \end{pmatrix}$$

Show that the matrix on the right is the inverse of the original matrix $\mathbf{Z} = \mathbf{A}^{-1}$ by multiplying it into $\mathbf{A}$. Now generate the solution set to the equations for which $\mathbf{A}$ is the coefficient matrix by multiplying $\mathbf{A}^{-1}$ into the nonhomogeneous vector to obtain $\{1, 2\}$.

Computer Projects

We shall use four BASIC programs to solve several problems involving up to seven simultaneous equations. The programs are: gaussian elimination (using a program called GAUSEL), Cramer's rule solution (using DET), the Gauss–Seidel method (using GSEID), and a matrix inversion method (using MINV; Chapter 3).

COMPUTER PROJECT 4-1 | Simultaneous Spectrophotometric Analysis

The first project in this chapter follows a problem given by Ewing (Ewing, 1985) in his widely used textbook on instrumental analysis. The problem is taken from the original research of Weissler (Weissler, 1945), who reacted hydrogen peroxide with Mo, Ti, and V ions in the same solution to produce compounds that absorb light strongly in overlapping peaks centered at 330, 410, and 460 nm, respectively (Fig. 4-2).

The absorbance A of a solution is given by Beer's law,

$$A = abc$$

where a is the absorptivity, a function of wavelength that is characteristic of the complex. The length of the light path through the absorbing solution is b in centimeters and c is the concentration of the absorbing species in grams per liter. If more than one complex is present, the

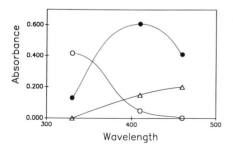

FIGURE 4-2 Visible absorption spectra of peroxide complexes of Mo, Ti, and V ions.

absorbance at any selected wavelength is the sum of contributions of each constituent.

Individual solutions of Mo, Ti, and V ion were complexed by hydrogen peroxide and each spectrum in the visible region was taken, with the results in Fig. 4-2. In each case, the absorbance of a single complex was recorded at the wavelength that gives the highest measured value for that component. The concentrations of the metal complex solutions were all the same: 40.0 mg/l. The absorbance table at λ_{max} constitutes a matrix with rows of absorbances, at one wavelength, of Mo, Ti, and V complex, in that order. Each column is comprised of absorbances for one metal complex at 330, 410, and 460 nm, in that order:

$$\begin{pmatrix} 0.416 & 0.130 & 0.000 \\ 0.048 & 0.608 & 0.148 \\ 0.002 & 0.410 & 0.200 \end{pmatrix}$$

Dividing each entry in the table by 0.04, to convert c to units of grams per liter, yields the absorptivity matrix in the data block of program GAUSEL (disk). Notice that the matrix has been arranged so that it is as nearly diagonal-dominant as the data set permits.

An unknown solution containing Mo, Ti, and V ions was treated with hydrogen peroxide and its absorbance was determined at the three wavelengths, in the same order (lowest to highest), that was used to generate the absorbance matrix for the single complexes. The ordered set of absorbances of any mixture of the complexes constitutes a B vector, in this case, $\mathbf{B} = \{0.284, 0.857, 0.718\}$. A set of simultaneous equations results. Taking M, T, and V to be the concentrations of the three metals involved, the unknown concentration vector $\mathbf{C} = \{M, T, V\}$ can be obtained by solving the simultaneous equation set

$$a_{1m}M + a_{1t}T \qquad\quad = 0.284$$
$$a_{2m}M + a_{2t}T + a_{2v}V = 0.857$$
$$a_{3m}M + a_{3t}T + a_{3v}V = 0.718$$

which is the same as the matrix equation

$$\mathbf{XC} = \mathbf{B}$$

where $\mathbf{X}$ is the absorptivity matrix, $\mathbf{B}$ is the nonhomogeneous vector, and $\mathbf{C}$ is the solution vector or concentration vector.

Because the matrix of coefficients does not change, the procedure can be used for any number of mixtures by observing the total absorbance at the three selected wavelengths, using the same absorptivity matrix to solve the simultaneous equation set.

Procedure Calculate the absorptivity matrix and, using program GAUSEL, solve the preceding equation set. Set up and solve the set resulting from a new set of observations leading to the absorbance vector

{0.327, 0.810, 0.673}. Record the approximate time necessary to solve each set.

Use the iterative program GSEID to generate the answer to three digits of self-consistency; that is, to arrive at a solution set for which one set of answers agrees with the previous one to within three digits. Do the same thing for five digits. How many iterations does this take?

COMPUTER PROJECT 4-2 | *Gauss–Seidel Iteration: Mass Spectroscopy*

The purpose of the project is to gain familiarity with the strengths and limitations of the Gauss–Seidel iterative method of solving simultaneous equations.

(a) Solve the prototypical equations

$$x + y = 3$$
$$x + 2y = 5$$

by the Gauss–Seidel iterative method. How many iterations are necessary to reach a self-consistent solution vector?

(b) Try to solve the set

$$x + 2y = 5$$
$$3x + y = 5$$

Reverse the order of the equations

$$3x + y = 5$$
$$x + 2y = 5$$

and try again. Comment on the results of this experiment in relation to the diagonal dominance of the coefficient matrix. Is the first set linearly dependent? Convergence of the Gauss–Seidel method is guaranteed if the sum of off-diagonal elements in each column is less than the diagonal element in that column.

(c) Solve the set

$$2x + y = 1$$
$$4x + 2.01y = 2$$

This set is said to be ill-conditioned because the second equation is almost an exact multiple of the first. The matrix of coefficients is almost singular.

(d) An example similar to Computer Project 4-1 involves quantitative analysis by mass spectrometry of four cyclic hydrocarbons (Isenhour and Jurs, 1985). The 4×4 matrix of sensitivity coefficients (analogous to absorptivities) has as its rows ethylcyclopenthane (Etcy), cyclohexane (Cy6), cycloheptane (Cy7), and methylcyclohexane (Mecy), at mass to charge (m/e) ratios of 69, 83, 84, and 98. Each m/e ratio constitutes a column; each entry in a given column represents the sensitivity coefficient for the designated hydrocarbon

	69	83	84	98
Etcy	121.0	9.35	1.38	20.2
Cy6	22.4	4.61	74.90	0.0
Cy7	27.1	20.7	1.30	32.8
Mecy	23.0	100.0	6.57	43.8

Procedure Solve this problem using GAUSEL, GSEID, DET, and MATINV (disk). Comment on the relative merits of the methods for this problem. Try rearranging the rows of the coefficient matrix and B vector to achieve diagonal dominance.

COMPUTER PROJECT 4-3 | Bond Enthalpies of Hydrocarbons

Derivation of bond enthalpies (energies) from thermochemical data involves some kind of system of simultaneous equations in which the sum of unknown bond enthalpies, each multiplied by the number of times that bond appears in a given molecule, is set equal to the enthalpy of atomization of that molecule (Atkins, 1986). Taking a number of molecules equal to the number of bond enthalpies one wishes to determine, one can generate an $n \times n$ set of equations in which the matrix of coefficients is populated by the (integral) number of bonds in the molecule, and the set of n atomization enthalpies is the B vector. (Obviously, each bond must appear at least once in the set.)

Carrying out this procedure for propane and butane yields the bond matrix and enthalpies of atomization in program BOND. Each enthalpy of atomization is obtained by subtracting the enthalpy of formation of the alkane from the sum of atomic atomization enthalpies (C: 715; H: 218 kJ mol^{-1}) for that molecule. For example, the molecular atomization enthalpy of propane is $3(715) + 8(218) - (-105) = 3994$ kJ mol^{-1}.

Procedure Run program BOND and interpret the results. Calculate the error vector, using a standard source of bond enthalpies (e.g., Roberts and Caserio, 1967; Atkins, 1986). Expand the method for 2-butene ($\Delta H_f = -11$ kJ mol^{-1}) and so obtain the C—H, C—C, and C=C bond enthalpies. These are approximate values only. There is a better way of doing this, as we shall see in Chapter 6.

COMPUTER PROJECT 4-4 | Balancing Redox Equations

The technique of balancing difficult redox equations using simultaneous equations of material balance has attracted a small but devoted following

(Bennett, 1954). The argument in its favor is that the oxidation number method of balancing chemical equations rests on a convenient fiction (the oxidation number itself) and the electron-transfer method depends on a process that may seem to the student, at least, to be equally fictional. Almost everyone believes in the equations of material balance, however.

An unbalanced redox equation, e.g.,

$$a\mathrm{Cr_2O_7^{2-}} + b\mathrm{H^+} + c\mathrm{Fe^{2+}} = d\mathrm{Cr^{3+}} + e\mathrm{H_2O} + f\mathrm{Fe^{3+}}$$

yields four fairly obvious algebraic equations: $c = f$, $2a = d$, $7a = e$, and $b = 2e$, which can be written $c - f = 0$, $2a - d = 0$, etc. and can be used to fill in the first four rows of a sparse six-column matrix. The equation of charge balance $-2a + b + 2c = 3d + 3f$ gives the fifth row in the coefficient matrix and arbitrarily setting $a = 1$, we obtain the sixth row, completing the $n \times n$ matrix and making the equation set nonhomogeneous. The program EQBAL generates the inverse matrix of the coefficients. Note how the inverse is related to the solution vector $\{1, 14, 6, 2, 7, 6\}$.

Procedure Modify EQBAL to balance the redox equation

$$\mathrm{Fe^{2+}} + \mathrm{MnO_4^-} + \mathrm{H^+} = \mathrm{Fe^{3+}} + \mathrm{Mn^{2+}} + \mathrm{H_2O}$$

Set up your matrix so that it is diagonal-dominant, i.e., so that the largest element in each row and column is on the principal diagonal. Go to the literature to find input data for a 7×7 matrix that can be used to solve a similar problem by Poages (Poages, 1945). Substitute the matrix and coefficient vector into program EQBAL and obtain the solution vector. Archive the new program with a unique file name, e.g., EQBAL1.

PROBLEMS | *Chapter 4*

1 Solve, by hand and using program GAUSEL,

$$x + y = 4$$
$$4x - 3y = -1.5$$

2 Solve

$$x + 2y - z + t = 2$$
$$x - 2y + z - 3t = 6$$
$$2x + y + 2z + t = -4$$
$$3x + 3y + z - 2t = 10$$

3 Solve, using Cramer's rule,

$$x \sin\theta + y \cos\theta = x'$$
$$-y \cos\theta + y \sin\theta = y'$$

4 Find the inverse of

$$\begin{pmatrix} -0.5 & 3/2 & 0 \\ -3/2 & -0.5 & 0 \\ 0 & 0 & 1 \end{pmatrix}$$

5 Balance the following equation by the method of Computer Project 4-4:

$$I^- + MnO_4^- + H_2O \rightarrow I_2 + MnO_2 + OH^-$$

6 Find the determinant

$$\det A = \begin{vmatrix} 2 & 1 & 0 \\ 1 & 0 & 1 \\ 3 & 3 & 2 \end{vmatrix}$$

7 Exchange any two columns of the determinant in Problem 6 and evaluate the new determinant. Exchange any two rows of Problem 6 and find the new determinant. Does the rule of Exercise 4-6 hold?

8 What is the atomization enthalpy of 2 moles of C and 3 moles of H_2? (See Computer Project 4-3.)

9 What is the atomization enthalpy of 1 mole of C_2H_6, ethane? The enthalpy of formation of ethane is -84 kJ mol^{-1}.

10 If the C—H bond enthalpy is 413 kJ mol^{-1}, what is the C—C bond enthalpy?

REFERENCES

Acton, F. S., 1970. *Numerical Methods that (Usually) Work*. Harper and Row, New York.

Anderson, J. M., 1966. *Mathematics for Quantum Chemistry*. Benjamin, New York, Section 3.2.

Atkins, P., 1986. *Physical Chemistry*, 3rd ed. W. H. Freeman, New York.

Bennett, G. W., 1954. *J. Chem. Educ.* **31**, 324.

Carnahan, B., Luther, H. A., and Wilkes, J. O., 1969. *Applied Numerical Methods*. Wiley, New York.

Ewing, G. W., 1985. *Instrumental Methods in Chemical Analysis*, 5th ed. McGraw-Hill, New York, Chapter 3.

Isenhour, T. L. and Jurs, P. C., 1979. *Introduction to Computer Programming for Chemists FORTRAN*, 2nd ed. Allyn and Bacon, Boston, MA.

Noggle, J. H., 1985. *Physical Chemistry on a Microcomputer*. Little, Brown and Co., Boston, MA.

Norris, A. C., 1981. *Computational Chemistry: An Introduction to Numerical Methods*. Wiley, New York, Chapter 5.

Poages, A., 1945. *J. Chem. Educ.*, **22**, 266.

Rice, J. R., 1983. *Numerical Methods, Software and Analysis*. McGraw-Hill, New York, Chapter 6.

Roberts, J. D. and Caserio, M. C., 1967. *Modern Organic Chemistry*. W. A. Benjamin, New York.

Scheid, F., 1968. *Numerical Analysis*. Schaum's Outline Series, McGraw-Hill, New York.

Weissler, A., 1945. *Ind. Eng. Chem., Anal. Ed.*, **17**, 695.

5 | *Curve Fitting*

A straight line drawn by eye through a scattered set of experimental points, presumed to represent a linear function, is not an acceptable representation of the data set. Computer programs exist that routinely fit data points with analytical functions, enabling us to avoid the subjectivity of visual curve fitting. Any set of experimental data points can be fitted more or less well by an analytical function. One must first select the function and then use routine curve fitting procedures to generate the analytical form of the function from the experimental observations. Then one can obtain statistical parameters that measure how good the fit is.

In this chapter, we shall use the principle of least squares to generate the equation of a unique curve for any given set of xy pairs of data points. The curve so obtained best fits the points subject to (1) assumption of an analytical form (straight line, quadratic, etc.) and (2) the assumption of randomly distributed deviations about the analytical form. We shall begin with the simplest case of a linear function passing through the origin to introduce the method and set up the ground rules. The more complicated cases of a linear function not passing through the origin will be solved by a method that is general. The method will be extended to nonlinear functions at the end of this chapter and elaborated in Chapter 6.

The Method of Least Squares

The probability function governing observation of outcome x_i from among a continuous random distribution of possible events x having a population mean μ and a population standard deviation σ is

$$p(x_i) \propto \exp\left(-(x_i - \mu)^2/2\sigma^2\right) \qquad (5\text{-}1)$$

Observing events according to a certain probability distribution requires that event x_1 (with probability p_1) occurs and x_2 (with probability p_2) occurs and so on. The probability of observing the entire distribution is the simultaneous probability of all events in the distribution occurring at once—that is, the product of the individual probabilities

$$p(x_1 \text{ and } x_2 \text{ and } \cdots) = p(x_1)p(x_2) \cdots = \prod p(x_i)$$

$$\prod p(x_i) \propto \prod \exp\left(-(x_i - \mu)^2/2\sigma^2\right) \tag{5-2}$$

By the nature of exponential numbers, $e^a e^b = e^{a+b}$,

$$\exp(x_1)\exp(x_2) \cdots = \exp(x_1 + x_2 + \cdots)$$

$$= \exp\left(\sum x_i\right)$$

hence

$$\prod p(x_i) \propto \exp\left(-\sum (x_i - \mu)^2/2\sigma^2\right) \tag{5-3}$$

Just as e^{-x} takes its maximum value when x is at a minimum, the right side of proportion (5-3) is a maximum when its exponent is a minimum. To minimize a fraction with a constant denominator, one minimizes the numerator

$$\sum (x_i - \mu)^2 = \text{a minimum} \tag{5-4}$$

In so doing, we obtain the maximum probability for the entire set of simultaneous events, i.e., the most probable distribution. The minimization condition [condition (5-4)] requires that the square of the differences between μ and all of the values x_i be as small as possible. It is reasonable to suppose that μ, subject to the minimization condition, will be the arithmetic mean, provided that the deviations are random, i.e., that the distribution is gaussian.

This method, because it involves minimizing the sum of squares of the deviations $x_i - \mu$, is called the *method of least squares*. It is very powerful and we shall use it in a number of different settings to obtain the best approximation to the data set of scalars (arithmetic mean), the best approximation to a straight line, and the best approximation to parabolic and higher-order data sets of two and more dimensions.

EXERCISE 5-1

For the simple data set $x_i = 2, 3, 7, 8, 10$ we have selected 5, 6, and 7 as possible values of μ. For which of these three is the sum of squared deviations from the data set a minimum?

Solution 5-1 The sum of squared deviations is least for $\mu = 6$. Conventional calculation of the arithmetic mean x shows that it is also 6.

Least Squares Minimization

Clearly, proposing arbitrary candidates for μ and selecting the one with the smallest value of $(x_i - \mu)^2$ is not very efficient, nor can it be readily generalized. A systematic method of arriving at the best value of μ is to find the minimum of $\Sigma(x_i - \mu)^2$ with respect to variation in μ. This is the point at which the first derivative is zero,

$$\frac{d}{d\mu} \Sigma (x_i - \mu)^2 = 0$$

The derivative is a sum of derivatives; hence,

$$\Sigma d/d\mu (x_i - \mu)^2 = -\Sigma 2(x_i - \mu) = 0$$

where μ is called a *minimization parameter* because the procedure amounts to selecting, from an infinite number of possible values, that μ for which the sum of squares of the deviations is at a minimum. Since μ is a constant summed over N terms,

$$\Sigma x_i = N\mu \tag{5-5a}$$

$$\mu = \frac{\Sigma x_i}{N} = \bar{x} \tag{5-5b}$$

which is the conclusion we reached earlier.

LINEAR FUNCTIONS PASSING THROUGH THE ORIGIN

If the linear function through the origin $y = mx$ were obeyed with perfect precision by an experimental data set (x_i, y_i), we would have

$$mx_i - y_i = 0$$

This is never the case for a real data set, which displays deviation d_i for each data point owing to experimental error. If the experimental error is random, the method of least squares applies to analysis of the set. For the real case,

$$mx_i - y_i = d_i \tag{5-6}$$

If we minimize the sum of squares of the deviations by differentiating with respect to m,

$$\frac{d}{dm} \Sigma (mx_i - y_i)^2 = 0 \tag{5-7}$$

which leads to

$$m = \frac{\Sigma x_i y_i}{\Sigma x_i^2} \tag{5-8}$$

The slope of the linear function is the minimization parameter. The only thing one can change to obtain a "better" fit to points for a line passing through the origin is its slope. The slope calculated by the least squares method is the "best" slope that can be obtained under the assumptions. Once one knows the slope of a linear function passing through the origin, one knows all that can be known about that function.

LEAST SQUARES FUNCTIONS NOT PASSING THROUGH THE ORIGIN

Deviations from a curve thought to be a straight line $y = mx + b$, not passing through the origin ($b \neq 0$), are

$$(mx_i + b) - y_i = d_i \qquad (5\text{-}9)$$

Note that m and b do not have subscripts because there is only one slope and one intercept; they are the minimization parameters for the least squares function.

Now there are two minimization conditions

$$\frac{\partial}{\partial b} \sum d_i^2 = 0$$

$$\frac{\partial}{\partial m} \sum d_i^2 = 0$$

which must be satisfied simultaneously. These are simultaneous equations. Substituting for d_i, one obtains

$$\sum mx_i + \sum b - \sum y_i = 0$$
$$\sum mx_i^2 + \sum bx_i - \sum x_i y_i = 0 \qquad (5\text{-}10)$$

which are called the *normal equations*. The normal equations can be solved by methods given in Chapter 4. Rewriting them as

$$Nb + \sum x_i m = \sum y_i$$
$$\sum x_i b + \sum x_i^2 m = \sum x_i y_i \qquad (5\text{-}11)$$

makes it clear that the intercept and slope are the two elements in the solution vector $\{b, m\}$ and the coefficient matrix and nonhomogeneous vector can be made up simply by taking sums of the experimental results or the sums of squares or products of results, all of which are available from the data set.

The coefficient matrix is symmetric

$$\begin{pmatrix} N & \sum x_i \\ \sum x_i & \sum x_i^2 \end{pmatrix} \qquad (5\text{-}12)$$

The nonhomogeneous vector is $\{\Sigma y_i, \Sigma x_i y_i\}$. Solving the normal equations by Cramer's rule (Exercise 4-5) leads to the solution set in determinantal form:

$$b = D_b/D$$

$$m = D_m/D \tag{5-13}$$

or

$$b = \frac{(\Sigma y_i)(\Sigma x_i^2) - (\Sigma x_i y_i)(\Sigma x_i)}{N\Sigma x_i^2 - (\Sigma x_i)^2}$$

$$m = \frac{N\Sigma x_i y_i - (\Sigma x_i)(\Sigma y_i)}{N\Sigma x_i^2 - (\Sigma x_i)^2} \tag{5-14}$$

which is the form usually given in elementary treatments of least squares data fitting. A general-purpose linear least squares curve-fitting program is included on the disk as LLSQ. Respond to the INPUT prompt with the number of data pairs.

Quadratic Functions

Many experimental functions approach linearity but are not really linear. (Many were historically thought to be linear until accurate experimental determinations showed some degree of nonlinearity.) Nearly linear behavior is often well represented by a quadratic equation

$$y = a + bx + cx^2 \tag{5-15}$$

The least squares derivation for quadratics is the same as it was for linear equations except that one more term is carried through the derivation and, of course, there are three normal equations rather than two. Random deviations from a quadratic are

$$\left(a + bx_i + cx_i^2\right) - y_i = d_i \tag{5-16}$$

The minimization conditions are

$$\frac{\partial d_i^2}{\partial a} = 0$$

$$\frac{\partial d_i^2}{\partial b} = 0 \tag{5-17}$$

$$\frac{\partial d_i^2}{\partial c} = 0$$

which must be true simultaneously. Solution of these equations leads to the normal equations

$$Na + b\sum x_i + c\sum x_i^2 = \sum y_i$$
$$a\sum x_i + b\sum x_i^2 + c\sum x_i^3 = \sum y_i x_i \qquad (5\text{-}18)$$
$$a\sum x_i^2 + b\sum x_i^3 + c\sum x_i^4 = \sum y_i x_i^2$$

with the solution vector $\{a, b, c\}$ and the nonhomogeneous vector $\{\sum y_i, \sum y_i x_i, \sum y_i x_i^2\}$. The matrix of the coefficients of the normal equations is

$$\begin{pmatrix} N & \sum x_i & \sum x_i^2 \\ \sum x_i & \sum x_i^2 & \sum x_i^3 \\ \sum x_i^2 & \sum x_i^3 & \sum x_i^4 \end{pmatrix} \qquad (5\text{-}19)$$

A general-purpose quadratic least squares program is included on the disk as QLSQ. Input format is the same as LLSQ.

Functions of Higher Degree

The form of the symmetric matrix of coefficients for the normal equations of the quadratic is very regular, suggesting a simple expansion to higher degree equations. The coefficient matrix for cubics is a 4×4 matrix with $\sum x_i^6$ in the 4, 4 position, the fourth degree has $\sum x_i^8$ in the 5, 5 position, and so on. It is routine (and tedious) to extend the method to higher degrees. Frequently, the experimental data are not sufficiently accurate to support higher degree calculations and "differences" in fit are artifacts of the calculation rather than features of the data set.

COMPUTER PROJECT 5-1 | *Linear Curve Fitting*

Linear extrapolation of the experimental behavior of a real gas to zero pressure or a solute to infinite dilution is often used as a technique to "get rid" of molecular or ionic interactions so as to study some property of the molecule or ion to which these interferences are considered extraneous. Emsley (1971) has studied the heat (enthalpy) of solution of potassium fluoride (KF) and the monosolvated species KF $\cdot$ HOAc in glacial acetic acid at several concentrations. A known weight of the anhydrous salt KF was added to a known weight of glacial acetic acid in a Dewar flask fitted with a heating coil, stirrer, and a sensitive thermometer; then the temperature change was recorded. The heat capacity C of the flask and its contents was determined by supplying a known amount of electrical energy Q to the flask and noting the temperature rise ΔT in kelvins:

$$Q(\text{joules}) = C\,\Delta T$$

The experiment was repeated for the solvated salt KF · HOAc, where the molecule of solvation is acetic acid, HOAc. Some experimental results [from a "reconstruction" of the experiment by Atkins (Atkins, 1986)] are as follows:

KF: $C = 4.168$ kJ K^{-1}

Molality	0.194	0.590	0.821	1.208
Temperature change, K	1.592	4.501	5.909	8.115

KF · HOAc: $C = 4.203$ kJ K^{-1}

Molality	0.280	0.504	0.910	1.190
Temperature change, K	−0.227	−0.432	−0.866	−1.189

Procedure Calculate the heats of solution of the two species, KF and KF · HOAc, at each of the four given molalities from a knowledge of the heat capacity. Calculate the enthalpy of solution per mole of solute $\Delta_{soln} H$ at each concentration. Use program LLSQ to determine the least squares curve fit for each species, the anhydrous and the solvated fluoride, to the function

$$\Delta_{soln} H = (SLOPE)M + \Delta_{soln} H^0$$

where M is the molality and $\Delta_{soln} H^0$ is the enthalpy of solution at infinite dilution. Determine the slopes of both functions SLOPE$_1$ and SLOPE$_2$. What are the units of SLOPE?

Read the article on the original research (Emsley, 1971) and include a commentary on these results in your report for this project.

COMPUTER PROJECT 5-2 | *The Boltzmann Constant*

An interesting historical application of the Boltzmann equation involves microscopic examination of the number density of very small spherical globules of latex suspended in water, which are distributed in the potential gradient of the gravitational field. If an arbitrary point in the suspension is selected, the number of particles N at height h μm (1 μm = 10^{-6} m) above the reference point can be counted. The number of particles per unit volume of the suspension in one series of measurements was

h/μm	0	50	70	90	100	150	200
N	977	453	293	219	176	69	28

where h is the height in micrometers above the reference point.

The Boltzmann distribution gives

$$N = N_0 \exp(-wgh/kT)$$

where w is the effective weight of the particle corrected for the buoyancy of the supporting medium. Taking logarithms,

$$\ln N = \ln N_0 - wgh/kT$$

and the slope of $\ln N$ vs. h is $-wg/kT$, where g is the acceleration due to the gravitational field. The supporting medium is water at 25°C ($\rho = 0.99727$) and the density of latex is 1.2049 g cm^{-3}. The latex particles had an average radius of 2.12×10^{-4} mm; hence, their effective weight was

$$w = v\rho = \left(\tfrac{4}{3}\right)\pi\left(2.12 \times 10^{-7} \text{ m}\right)^3(1.2049 - 0.99727)$$

where the last term is a correction for buoyancy. This yields $m' = 8.287 \times 10^{-18}$ kg.

Procedure Compute the slope of the function by a linear least squares procedure and obtain a value of Boltzmann's constant. How many particles do you expect to find 125 μm above the reference point?

Let us suppose that the uncertainty in counting the particle distribution is so much larger than all other error sources that they may be ignored and that the uncertainty can be expressed as twice the standard deviation in $\ln N$ about the linear function $\ln N$ vs. h. Is the modern value of 1.381×10^{-23} within these error limits?

Having determined k and knowing R from macroscopic measurements on gases, determine Avogadro's number L from the relationship

$$R = kL$$

Calculate the percent difference between L found by this method and the modern value of 6.022×10^{23}.

COMPUTER PROJECT 5-3 | *The Ionization Energy of Hydrogen*

The ionization energy for hydrogen is the minimum amount of energy that is required to bring about the reaction

$$\text{H} \rightarrow \text{H}^+ + e^-$$

The ionization energy for hydrogen (or other hydrogen-like systems) can be determined using the Rydberg equation

$$\nu = \frac{1}{\lambda} = R\left(\frac{1}{n_1^2} - \frac{1}{n_2^2}\right)$$

along with an accurate set of spectral data. This leads us to believe that both the slope and intercept of the linear function

$$\nu = R - R\left(\frac{1}{n_2}\right)$$

for ground-state hydrogen atoms ($n_1 = 1$) are R, because the ionization energy has $n_2 = \infty$.

Procedure Use LLSQ to determine a value of the ionization energy of hydrogen from the following wave numbers taken from the Lyman series of the hydrogen spectrum ($n_1 = 1$).
Spectral wavenumbers ν cm^{-1}

| 82259 | 97492 | 102824 | 105292 | 106632 | 107440 |

Use the results of this data treatment to obtain a value of the Rydberg constant R. Compare the value you obtain with an accepted value. Quote the source of the accepted value you use for comparison in your report. What are the units of R? A conversion factor may be necessary to obtain unit consistency. Express your value for the ionization energy of H in units of electron volts (eV). We will need it later (Chapter 8).

COMPUTER PROJECT 5-4| *The Partial Molal Volume of ZnCl$_2$*

In general, the volume of a solution, say $ZnCl_2$ in water, is dependent on the number of moles n_i of each of the components. For a binary solution,

$$V = f(n_1, n_2)$$

The change in volume dV upon adding a small amount dn_1 of water or dn_2 of $ZnCl_2$ is

$$dV = \left(\frac{\partial V}{\partial n_1}\right)_{n_2} dn_1 + \left(\frac{\partial V}{\partial n_2}\right)_{n_1} dn_2$$

where we stipulate that P and T are constant for the process and we adopt the usual subscript convention: 1 for solvent and 2 for solute. If we specify 1 kg as the amount of water, the number of moles of $ZnCl_2$ or any other solute n_2 is the *molality* of that solute. We expect that the volume of the solution will be greater than 1000 cm^3 by the volume taken up by the $ZnCl_2$. It may seem reasonable to take the volume of 1 mol of $ZnCl_2$ in the solid state V_m and add to 1000 cm^3 to get the volume of a 1-molal solution. One-half the molar volume of solute would, by this scheme, lead to the volume of a 0.5-molal solution and so on. This does not work. The volumes are not additive. Indeed, a few solutes cause contraction of the solution to less than 1000 cm^3.

Interactions at the molecular or ionic level cause an expansion or contraction of the solution so that, in general,

$$V \neq 1000 + V_m$$

We define a partial molar volume $\overline{V}_i$ such that

$$V = n_1 \overline{V}_1 + n_2 \overline{V}_2$$

It can be shown (Alberty, 1987) that

$$\overline{V}_i = \left(\frac{\partial V}{\partial n_i} \right)_{n_j}$$

where the subscript n_j indicates that all components in the solution other than i are held constant. If the solution is a binary solution of n_2 mol of solute in 1 kg of water, $\overline{V}_2$ is the partial *molal* volume of component 2. A partial molal volume is a special case of the partial molar volume for 1 kg of solvent (see Problem 8-1, Atkins, 1986).

Procedure A study on the partial molal volume of $ZnCl_2$ solutions gave the following data (Alberty, 1987):

% by weight	2	6	10	14	18	20
Density	1.0167	1.0532	1.0891	1.1275	1.1665	1.1866

Calculate the number of moles of $ZnCl_2$ per kilogram of water in each solution (the molality m). Calculate the volume V of solution containing 1 kg of water at each solute concentration. Plot V vs. m. Use program QLSQ to obtain a quadratic expression $V = f(m)$. Obtain an expression for the slope dV/dm, which is the same as $(\partial V / \partial n_2)_{n_1}$. This is the partial molal volume of $ZnCl_2$. It is a partial because V varies with both n_{ZnCl_2} and the number of moles of water n_1. What is the partial molar volume of $ZnCl_2$ in water at 1-molal concentration? What is the partial molal volume of water at this concentration?

PROBLEMS | Chapter 5

1 Select several values for μ of the data set in Exercise 5-1 and calculate $x_i - \mu$ for each of them. Plot the curve of $\Sigma(x_i - \mu)^2$ as a function of the selected parameter μ and visually locate the minimum. Compare with Solution 5-1.

2 Obtain Eq. (5-8) from Eq. (5-6) *via* Eq. (5-7).

3 An excess of porphobilinogen in the urine is associated with hepatic disorders and lead poisoning. Porphobilinogen can be separated from other porphyrins by ion exchange chromatography and treated with *p*-dimethylaminobenzaldehyde (PDMA) to produce a red compound that absorbs light strongly at 550 nm. A set of standard solutions was made up with concentrations of 50.0, 75.0, 100.0, 125.0, 150.0, 175.0, 200.0, 225.0, and 250.0 mg/100 mL of porphobilinogen. Their absorbances A after treatment with PDMA were 0.039, 0.061, 0.087, 0.107, 0.119, 0.163, 0.179, 0.194, and 0.213. What is the spectrophotometric calibration curve $A = f$(concentration) for this method? What are the units of slope? Three urine specimens treated by this method yielded absorbances A of 0.180, 0.162, and 0.213. What were the porphobilinogen concentrations of these three samples?

4 Expand the three determinants D, D_b, and D_m for the least squares fit to a linear function not passing through the origin so as to obtain explicit algebraic expressions for b and m, the y intercept, and the slope of the best straight line representing the experimental data.

5 Set up the determinants D, D_a, D_b, and D_c for the least squares fit to a parabolic curve not passing through the origin.

6 Expand the four 3×3 determinants obtained in Problem 5.

7 Using the expanded determinants from Problem 6, write explicit algebraic expressions for the three minimization parameters a, b, and c for a parabolic curve fit.

8 Compare the solution for Problem 7 with the BASIC statements in the program QLSQ. They should agree.

9 Write the determinant for a sixth degree curve fitting procedure.

10 The volume of $ZnCl_2$ solutions containing 1000 g of water varies according to the quadratic equation (Computer Project 5-4)

$$V = 4.471 \text{ m}^2 + 21.148 \text{ m} + 999.71$$

Find the partial molal volume of $ZnCl_2$ in these solutions at 0.5-, 1.0-, 1.5-, and 2.0-molar concentrations.

REFERENCES

Alberty, R. A., 1987. *Physical Chemistry*, 7th ed. Wiley, New York.

Atkins, P. W., 1986. *Physical Chemistry*, 3rd ed. W. H. Freeman, New York.

Barrante, J. R., 1974. *Applied Mathematics for Physical Chemistry*. Prentice-Hall, Englewood Cliffs, NJ.

Chatterjee, S. and Price, B., 1977. *Regression Analysis by Example*. Wiley, New York.

Emsley, J., 1971. *J. Chem. Soc.* p. 2702.

Rogers, D. W., 1983. *BASIC Microcomputing and Biostatistics*. Humana, Clifton, NJ.

Young, H. D., 1962. *Statistical Treatment of Experimental Data*. McGraw-Hill, New York.

6 | *Multivariate Least Squares Analysis*

In this chapter, we shall treat the multivariate case of least squares analysis, i.e., the case for which the dependent variable is a function of two or more independent variables. Because matrices are so conveniently handled by computer and because the mathematical formalism is simpler, multivariate analysis will be developed as a topic in matrix algebra rather than conventional algebra. Applications to experimental data from analytical chemistry and thermodynamics will be given. Bond enthalpies are discussed in preparation for resonance energy determination in Chapter 8 and in relation to enthalpy computations in Chapters 10 through 13.

The univariate least squares regression procedure (Chapter 5) arrives at the "best" values of the parameters m and b in the equation $y = mx + b$ (or higher degree) by minimizing the squares of the deviations of experimental observations y_i from the curve defined by m and b (or $a, b, c, \ldots$). To simplify the algebra, the error in the x variable can be considered negligible relative to that of the y variable, though this is not a necessary condition.

We have already seen the normal equations in matrix form. In the multivariate case, there are as many slope parameters as there are independent variables and there is one intercept. The simplest multivariate problem is that in which there are only two independent variables and the intercept is 0:

$$y = m_1 x_1 + m_2 x_2$$

The function describes a plane passing through the origin. Let us restrict it to positive values of x_1, x_2, and y. One measurement of the dependent variable yields y_1 for known values of the independent variables x_{11}, x_{12},

$$y_1 = m_1 x_{11} + m_2 x_{12} \tag{6-1a}$$

and a second yields

$$y_2 = m_1 x_{21} + m_2 x_{22} \tag{6-1b}$$

for new values of the independent variables x_{21}, x_{22}.

The mathematical requirements for unique determination of the two slopes m_1 and m_2 are satisfied by these two measurements, provided that the second equation is not a linear combination of the first. In practice, however, because of experimental error, this is a minimum requirement and may be expected to yield a least-reliable solution set for the system, just as establishing the slope of a straight line through the origin by one experimental point may be expected to yield a least-reliable slope, inferior in this respect to the slope obtained from 2, 3, or p experimental points. In univariate problems, accepted practice dictates that we obtain many experimental points and determine the best slope representing them by a suitable univariate regression procedure.

The analogous procedure for a multivariate problem is to obtain many experimental equations and to extract the best slopes from them by regression. Optimal solution for n unknowns requires that the slope vector be obtained from p equations, where p is larger than n, preferably much larger. When there are more than the minimum number of equations from which the slope vector is to be extracted, we shall say that the equation set is an *overdetermined* set. Clearly, n equations can be selected from among the p available equations, but this is precisely what we do not wish to do because we must subjectively discard experimental data that may have been gained at considerable expense in time and money.

Equations (6-1) can be written in matrix form

$$Y = MX \qquad (6\text{-}2)$$

where X is the matrix of (known) input variables

$$X = \begin{pmatrix} x_{11} & x_{12} \\ x_{21} & x_{22} \end{pmatrix} \qquad (6\text{-}3)$$

Y is the nonhomogeneous vector of dependent experimental measurements, and M is the slope vector, i.e., the solution vector of the regression problem.

The deviations or residuals of y_1 and y_2 from the regression plane passing through the origin are

$$d_1 = m_1 x_{11} + m_2 x_{12} - y_1 \qquad (6\text{-}4a)$$

$$d_2 = m_1 x_{21} + m_2 x_{22} - y_2 \qquad (6\text{-}4b)$$

which are minimized by setting $\partial/\partial m_1 \, \Sigma d_i^2$ and $\partial/\partial m_2 \, \Sigma d_i^2$ equal to zero as shown in Chapter 5. These yield

$$m_1 x_{11}^2 + m_2 x_{11} x_{12} + m_1 x_{21}^2 + m_2 x_{21} x_{22} = y_1 x_{11} + y_2 x_{21} \qquad (6\text{-}5a)$$

$$m_1 x_{11} x_{12} + m_2 x_{12}^2 + m_1 x_{21} x_{22} + m_2 x_{22}^2 = y_1 x_{12} + y_2 x_{22} \qquad (6\text{-}5b)$$

We are dealing with real numbers that commute; hence, it is evident that the right side of Eqs. (6-5) is

$$\begin{pmatrix} x_{11} & x_{21} \\ x_{12} & x_{22} \end{pmatrix} \begin{pmatrix} y_1 \\ y_2 \end{pmatrix}$$

or

$$\mathbf{X}'\mathbf{Y}$$

where the matrix in x is the transpose of the input matrix $\mathbf{X}$.

The simple matrix expression $\mathbf{X}'\mathbf{Y}$ represents the right side of the normal equations for this or any larger set. The left side of the normal equations can be seen to include $\mathbf{X}$, its transpose, and $\mathbf{M}$ as a product. Matrix multiplication shows that

$$\mathbf{X}'\mathbf{X}\mathbf{M}$$

is the matrix representation of the left side of the normal equations (see Problems).

Two important facts emerge here. First, the method is general and can be worked out for the $n \times n$ case ($n > 2$) as before, but with added labor. Second, *the input matrix need not be square*. By the geometric nature of a rectangular matrix, it is always conformable for multiplication into its own inverse. The result is a square product matrix of the smaller of the two dimensions of the rectangular input matrix. Indeed, for the present treatment to be nontrivial, the input matrix must be rectangular; a square input matrix with $\mathbf{X}^{-1}$ defined (X nonsingular) represents the problem of n independent equations in n unknowns, which is the problem we said we do not want to solve. From this point on, envision $\mathbf{X}$ as a $p \times n$ matrix with $p > n$.

The normal equations are simultaneous equations

$$\mathbf{X}'\mathbf{X}\mathbf{M} = \mathbf{X}'\mathbf{Y} \tag{6-6}$$

in which $\mathbf{X}'\mathbf{Y}$ yields a vector of the smaller dimension of $\mathbf{X}'$, the same dimension as the vector obtained as the product $\mathbf{X}'\mathbf{X}\mathbf{M}$. The normal equations are often written

$$(\mathbf{X}'\mathbf{X})\mathbf{M} = \mathbf{Q} \tag{6-7}$$

where $\mathbf{X}'\mathbf{X}$ emerges as the coefficient matrix of a simple set of simultaneous equations, $\mathbf{M}$ is the solution vector, and $\mathbf{Q}$ is the nonhomogeneous vector $\mathbf{X}'\mathbf{Y}$. Solution follows by the usual method of inverting the coefficient matrix and premultiplying it into both sides

$$(\mathbf{X}'\mathbf{X})^{-1}(\mathbf{X}'\mathbf{X})\mathbf{M} = (\mathbf{X}'\mathbf{X})^{-1}\mathbf{Q}$$

or

$$\tag{6-8}$$

$$\mathbf{M} = (\mathbf{X}'\mathbf{X})^{-1}\mathbf{Q}$$

This equation permits us to generate an n-fold slope vector $\mathbf{M}$ from a rectangular matrix $\mathbf{X}$ and a dependent variable vector $\mathbf{Y}$. The procedure is analogous to the univariate case of generating the slope m of a calibration curve, passing through the origin, from p known values of x and the corresponding measured values $y_1, \ldots, y_i, \ldots, y_p$, with the intention of using m to determine unknown values of x (the concentration of an analyte perhaps) from future measurements of y. In the multivariate case, one slope vector is not enough; a square slope matrix must be generated with dimensions equal to the number of unknowns one wishes to determine,

$$\begin{pmatrix} m_{11} & m_{12} \\ m_{21} & m_{22} \end{pmatrix} \tag{6-9}$$

for a problem in two unknowns, and larger for n unknowns. This is done by repeating the procedure just given n times with different vectors $\mathbf{Y}_1, \ldots, \mathbf{Y}_i, \ldots, \mathbf{Y}_p$ to produce an $n \times n$ slope matrix $\mathbf{M}$ with the intention of using $\mathbf{M}$ to determine unknown values of $\mathbf{X}$,

$$\mathbf{X} = \mathbf{M}^{-1}\mathbf{Y} \tag{6-10}$$

The n-fold procedure $(n > 2)$ produces an n-dimensional hyperplane in $(n + 1)$-dimensional space. Lest this seem unnecessarily abstract, we may regard the $n \times n$ slope matrix as the matrix establishing a calibration surface from which we may determine n unknowns x_j by making n independent measurements y_i. An illustrative example generates a 2×2 calibration matrix from which we can determine the concentrations x_1 and x_2 of dichromate and permanganate ions by making spectrophotometric measurements y_1 and y_2 on an aqueous mixture of the unknowns.

Application: Simultaneous Analysis by Visual Spectrophotometry

The problem of $Cr_2O_7^{2-}$ and MnO_4^- determination is well known (Computer Project 4-1; Ewing, 1985). It requires determination of a matrix of four calibration constants, one for each unknown concentration at each wavelength, where there are as many wavelengths as unknowns in the equation set

$$\begin{aligned} a_{11}x_1 + a_{12}x_2 &= A_1 \\ a_{21}x_1 + a_{22}x_2 &= A_2 \end{aligned} \tag{6-11}$$

for the two-dimensional case. The elements a_{ij} are absorptivities (or are proportional to absorptivities, depending on the concentration units and cell dimensions), $\mathbf{X}$ is the unknown concentration vector, and $\tilde{\mathbf{A}} = \{A_1, A_2\}$

is the absorbance vector, observed at wavelength λ_i:

$$\mathbf{AX} = \tilde{\mathbf{A}} \tag{6-12}$$

(For accepted nomenclature, see Ewing, 1985.)

As a practical note, the wavelengths λ_i should be chosen so as to make the a_{ij} matrix as "nonsingular as possible," i.e., at each wavelength, absorbance by one species should dominate, insofar as possible, all the rest.

Clearly, dichromate–permanganate determination is an artificial problem, because the matrix of coefficients can be obtained from four univariate least squares regression treatments, one on $Cr_2O_7^{2-}$ at 440 nm, one at 525 nm, and one on MnO_4^- at each of these wavelengths. We did this using five concentrations of each absorbing species and obtained the matrix

$$\begin{pmatrix} 4.39 \times 10^{-3} & 2.98 \times 10^{-3} \\ 3.85 \times 10^{-4} & 4.24 \times 10^{-2} \end{pmatrix} \tag{6-13}$$

Elements in the slope matrix are proportional to absorptivities and concentrations are in parts per million.

Applying program MULTVAR with the input concentration matrix

$$\begin{pmatrix} 53.0 & 8.65 \\ 27.0 & 13.0 \\ 80.0 & 4.33 \\ 0.0 & 17.3 \\ 106 & 0.0 \end{pmatrix} \tag{6-14}$$

and the measured vector of absorbances $\mathbf{A}_j = \{0.251, 0.149, 0.361, 0.049, 0.456\}$ at 440 nm, we obtained the a_{1j} vector $\{4.32 \times 10^{-3}, 2.72 \times 10^{-3}\}$ rounded to the appropriate number of significant figures. This is the top row of the slope matrix; run time is less than 4 s.

Repeating the process with a new measured vector, $\{0.401, 0.568, 0.209, 0.740, 0.042\}$ at 525 nm leads to the a_{2j} vector $\{3.85 \times 10^{-4}, 4.29 \times 10^{-2}\}$, the bottom row of the slope matrix. Subtracting the slope matrix obtained by the multivariate least squares treatment from that obtained by univariate least squares yields the error matrix

$$\begin{pmatrix} 0.07 \times 10^{-3} & 0.26 \times 10^{-3} \\ 0.0 & -0.50 \times 10^{-3} \end{pmatrix}$$

where the univariate results are taken as "true" values.

Error Analysis

Normally, one does not have "true" values of the elements of the slope matrix $\mathbf{M}$ for comparison. It is always possible, however, to obtain $\mathbf{YS}$, the vector of predicted y values at each of the known x_i from the slope vector **M**:

$$YS = MX \tag{6-15}$$

This leads to a vector of residuals

$$E = YS - Y = MX - Y$$

The best estimator (Chatterjee and Price, 1977) of the variance is

$$s^2 = E^t E / (p - n - 1) \tag{6-16}$$

Under the assumption that the residuals are normally distributed, the best estimator of the variance of the ith element in **M** is $s^2 C_{ii}$ where C_{ii} is the ith diagonal element of $C = (X^t X)^{-1}$,

$$C = \begin{pmatrix} 5.32 \times 10^{-5} & -1.09 \times 10^{-4} \\ -1.09 \times 10^{-4} & 2.01 \times 10^{-3} \end{pmatrix}$$

for this data set.

Following these equations for the data set at 440 nm,

$$E = \{-0.001, -0.003, 0.004, 0.002, -0.002\}$$

which leads to a sample variance $s^2 = 1.7 \times 10^{-5}$. Student's t tests of various null hypotheses (Rogers, 1983) follow in the usual way as do the 95% confidence limits on the computed slopes, in particular, $\{(4.32 \pm 0.13) \times 10^{-3}, (2.72 \pm 0.79) \times 10^{-3}\}$ at 440 nm and $\{(3.85 \pm 2.7) \times 10^{-4}, (4.29 \pm 0.16) \times 10^{-2}\}$ at 525 nm. The relative uncertainty on element a_{21} is large because the parameter is an order of magnitude smaller than the other elements in the slope matrix.

COMPUTER PROJECT 6-1 | *Calibration Surfaces Not Passing through the Origin*

If the generalization of Eqs. (6-1),

$$y_i = \sum m_j x_{ij}$$

contains a term, call it m_0, with the stipulation that x_{i0} is always 1, the normal equations and the solution for the M vector follow just the same as they did in the previous section except that each equation in Eqs. (6-4) contains an additive constant m_0. The constant m_0, a minimization parameter, along with the rest of the m_j is the best estimator of the y intercept for a function not passing through the origin; it is the unique point at which the calibration surface cuts the y axis.

To set up the problem for a microcomputer, one need only enter the input matrix with a 1 as each element of the 0th or leftmost column. Suitable modifications must be made to accommodate matrices larger in one dimension than the X matrix of input data and output vectors containing one more minimization parameter than before, the intercept m_0.

Procedure The new method can be tested using the matrix of concentrations, in micromoles per liter, of tryptophan and tyrosine at 280 nm, suitably modified to take into account constant absorption at 280 nm of some absorber that is neither tryptophan nor tyrosine:

$$\begin{pmatrix} 1 & 47.6 & 116 \\ 1 & 125 & 147 \\ 1 & 23.7 & 109 \\ 1 & 156 & 48.3 \\ 1 & 272 & 15.1 \end{pmatrix}$$

Columns 2 and 3 of the concentration matrix are analogous to the concentration matrix Eq. (6-14). Using program MULTVAR with the absorbance vector {0.846, 1.121, 0.776, 0.599, 0.559}, calculate the solution vector. The first element of the solution vector is the intercept due to background absorption and the second two elements are the absorbancies. What are the absorbancies of tyrosine and tryptophan at 280 nm by this method? Compare your results with accepted values (Eisenberg and Crothers, 1979). What is the intercept of absorbance due to compounds of the solution that are neither tyrosine nor tryptophan?

COMPUTER PROJECT 6-2 | *Bond Energies of Hydrocarbons*

Determination of bond energies in hydrocarbons is a nontrivial example of multivariate analysis because lone C—C and C—H bonds cannot be observed; they always appear in groups. One could take the C—H bond energy to be one-fourth of the energy of atomization of methane, subtract six times that value from the energy of atomization of ethane to get the C—C single bond energy, and proceed in a like way using the atomization energies of an alkene and an alkyne to generate the carbon–carbon double and triple bonds. This strategy would be risky at best because the C—H bonds in a single molecule (methane) would be taken to represent all C—H bonds, ethane would be taken to represent all C—C bonds, and so on. It would be better to draw bond energies from a *basis set* of data for several, preferably many, molecules on the reasonable assumption that the mean or average result is more likely to be right than any single result from the set. Because the bonds cannot be observed singly, the problem is multivariate and because we wish to generate a few bond energies from many experimental results, the input matrix will be overdetermined.

We shall generate the energies for the carbon–hydrogen bond B_{CH} and the carbon–carbon single bond B_{CC} using the five linear alkanes from ethane through hexane as the five-member data base. The equation used is

$$hB_{CH} + sB_{CC} = \Delta H_a$$

where h is the number of C—H bonds in each hydrocarbon, s is the corresponding number of C—C single bonds, and ΔH_a is the enthalpy of atomization. Enthalpies of atomization of carbon and hydrogen were taken as 715 and 218 kJ mol^{-1} of atoms produced (Lewis et al., 1961) and were combined with the appropriate enthalpy of formation (Cox and Pilcher, 1970) to obtain the enthalpy of atomization of each hydrocarbon.

Procedure Decide upon an appropriate input matrix of bond numbers

$$\begin{pmatrix} h_1 & s_1 \\ h_2 & s_2 \\ \vdots & \vdots \\ h_n & s_n \end{pmatrix}$$

for ethane through hexane. The enthalpy of atomization vector is $\{2823, 3994, 5166, 6338, 7509\}$ in the same order. Compute the two-fold vector of bond enthalpies. Obtain an error vector by comparing your result with the accepted values (Atkins, 1986) of 413 and 346 for the C—H and C—C bonds, respectively.

More than a reasonable number of significant figures are produced by the calculation. When properly rounded, the uncertainty of the computed result should be reflected in the significant figures such that the rightmost digit is uncertain, but no more than one uncertain digit is included in the final result. This is, of course, an approximate indicator of uncertainty. If a rigorous indicator is desired, the standard deviation, variance, or confidence level should be reported with the computed result.

COMPUTER PROJECT 6-3| *Expanding the Basis Set*

Add ethylene, propene, 1-butene, acetylene, and propyne to the basis set. To do this, you must calculate five new atomization enthalpies (from a cycle similar to the Born–Haber cycle). Also extend the input matrix to a 4×10 matrix. Generate the C—H, C—C, C=C, and C≡C bond energies. Comment on the magnitude of the bond enthalpies, particularly the enthalpies of the C—C single, double, and triple bonds. Look up a set of accepted values for these bond energies and calculate a four-fold error vector. Does the error for C—H and C—C get larger or smaller for the extended basis set as compared with the smaller basis set used in Computer Projects 4-3 and 6-2? Discuss this result.

PROBLEMS | *Chapter 6*

1 Obtain the normal equations [Eqs. (6)] from the minimization conditions $\partial \Sigma d_i^2 / \partial m_i = \partial \Sigma d_i^2 / \partial m_2 = 0$.

2 Multiply $\mathbf{X'Y}$ from Eq. (6-6) to show that it is equal to the right side of Eq. (6-5).

3 Multiply $\mathbf{X'XM}$ from Eq. (6-6) to show that it is equal to the left side of Eq. (6-5).

4 Can a rectangular matrix be both premultiplied or postmultiplied into its own transpose, or must multiplication be either pre- or post- for conformability? If it must be one or the other, which is conformable?

5 What is the average enthalpy of atomization of the four $C—H$ bonds in methane? Compare this value with the accepted value of the $C—H$ bond enthalpy.

6 Calculate the bond enthalpy of the $C—C$ bond in ethane using only the enthalpies of atomization of methane and ethane. Compare this result with the accepted result.

7 When 6 moles of the substrate analog PALA combine with the enzyme ACTase, two things happen at the same time: The enzyme T unfolds to a more active form R,

$$T \rightarrow R$$

and 6 moles of PALA bind to the enzyme. The measured enthalpy of reactions is

$$\Delta H = 6 \, \Delta H_{PALA} + \Delta H_{T \rightarrow R} = -209.2 \text{ kJ mol}^{-1}$$

(Klotz and Rosenberg, 1986).

We would like to know the binding enthalpy per mole of PALA and the enthalpy of transformation for $T \rightarrow R$, but we do not have enough information. Independent studies show that partial unfolding of ACTase occurs upon binding of less PALA, in particular, that 1.8 mole of PALA cause 43% unfolding and 4.8 moles causes 86%. The enthalpy changes are -63.2 and -184.5 kJ mol^{-1}, respectively, leading to

$$1.8 \, \Delta H_{PALA} + 0.43 \, \Delta H_{T \rightarrow R} = -63.2$$

and

$$4.8 \, \Delta H_{PALA} + 0.86 \, \Delta H_{T \rightarrow R} = -184.5$$

both in kilojoules per mole. Use all three equations with MLTVAR to find the enthalpies of binding and of unfolding for this enzyme.

REFERENCES

Atkins, P. W., 1986. *Physical Chemistry*, 3rd ed. W. H. Freeman, New York.

Chatterjee, S. and Price, B., 1977. *Regression Analysis by Example*. Wiley, New York.

Cox, J. D. and Pilcher, G., 1970. *Thermochemistry of Organic and Organometallic Compounds*. Academic, London.

Eisenberg, D. S. and Crothers, D. M., 1979. *Physical Chemistry with Applications to the Life Sciences*. Benjamin/Cummings, Menlo Park, CA.

Ewing, G. W., 1985. *Instrumental Methods of Chemical Analysis*, 5th ed. McGraw-Hill, New York.

Klotz, I. M. and Rosenberg, R. M., 1986. *Chemical Thermodynamics*. Benjamin/Cummings, Menlo Park, CA.

Lewis, G. N., Randall, M., Pitzer, K. S., and Brewer, L., 1961. *Thermodynamics*, 2nd ed. McGraw-Hill, New York.

Rogers, D. W., 1983. *BASIC Microcomputing and Biostatistics*. Humana Press, Clifton, NJ.

7 | Molecular Orbital Calculations I

There are few problems in chemistry (none, according to Dirac) that could not be solved if we had a general method of obtaining exact solutions of the Schroedinger equation

$$H\psi = E\psi \tag{7-1}$$

Mathematically, this appears to be unattainable. We must be satisfied with approximate solutions to the Schroedinger equation, although ever-improving approximations are being obtained by modern computational methods.

The simplest method to use and the one involving the most drastic approximations and assumptions is the Huckel method. The great strength of the Huckel method is that it provides a semiquantitative theoretical treatment of ground-state energies, bond orders, electron densities, and free valences that appeals to the pictorial sense of molecular structure and reactive affinity that most chemists use in their everyday work.

EXACT SOLUTIONS OF THE SCHROEDINGER EQUATION

Among the few systems that can be solved exactly are the particle in a one-dimensional "box" (see Fig. 7-1), the hydrogen atom, and the hydrogen molecule ion H_2^+. Although of limited interest chemically, these systems serve as models for more general, chemically important approximate methods.

The hamiltonian H is an operator, that is, it is an instruction telling you what operation or operations to perform on the wave function ψ. The hamiltonian takes a different form for different mechanical systems, which is why a single equation can be said to represent very different kinds of mechanical motion: electrons in chemical bonds, vibrating molecules, torsional rotations of methyl groups, and so on. One of the simplest forms of H is

$$H = -\frac{\hbar^2}{2m} d^2/dx^2 + V$$

81

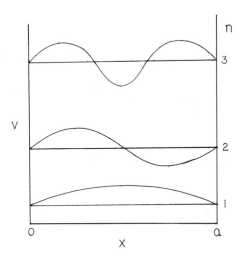

FIGURE 7-1 Energy levels and nodal properties of the wave function for a particle in a one-dimensional potential well.

where V, the potential energy, is zero for excursions of x between two limits, designated 0 and a, but is infinite elsewhere. This form leads to the Schroedinger equation for a particle in a one-dimensional box,

$$\left[-\frac{\hbar^2}{2m} d^2/dx^2 + V \right] \psi = E\psi$$

where $\hbar = h/2\pi$. This problem is treated in elementary physical chemistry books (e.g., Atkins, 1986).

The hydrogen atom is a three-dimensional problem in which the attractive force of the nucleus has spherical symmetry. Therefore, it is advantageous to set up and solve the problem in spherical polar coordinates r, θ, and ϕ. The resulting equation can be broken up into three parts, one a function of r only, one a function of θ only, and one a function of ϕ. These can be solved separately and exactly. Each equation leads to a quantum number:

$$R(r) \rightarrow n$$
$$\Theta(\theta) \rightarrow l$$
$$\Phi(\phi) \rightarrow m$$

These are three of the four quantum numbers familiar from general chemistry. The spin quantum number s arises when relativity is included in the problem.

The hydrogen molecule ion is set up best in confocal elliptical coordinates with the two protons at the foci of the ellipse and one electron moving in their combined potential field. Solution follows in much the same way as it did for the hydrogen atom. Solution is exact for this system (Hanna, 1981).

In the few cases that permit exact solution of the Schroedinger equation, the energy of the system is the sum of energies obtained by solving the separated equations and the wave function is the product of the wave functions obtained for the separated equations.

APPROXIMATE SOLUTIONS

We have said that the Schroedinger equation for molecules cannot be solved exactly. This is because the exact equation is not separable. One strategy is to make assumptions that permit us to write approximate forms of the Schroedinger equation for molecules and ultimately to separate the new and approximate equations. There is then a choice as to how to solve the separated equations. The Huckel method is one possibility. The self-consistent field method (Chapter 9) is another.

Three major approximations are made to separate the Schroedinger equation into a set of smaller equations.

Born–Oppenheimer Approximation

We assume that the nuclei are so slow-moving relative to electrons that we may regard them as fixed masses. This amounts to separation of the Schroedinger equation into two parts, one for nuclei and one for electrons. We then drop the nuclear kinetic energy operator and fix the internuclear repulsion terms from the potential, but we retain all terms that involve electrons, including the potential energy of attractive forces between nuclei and electrons. In chemical calculations, we usually write the Schroedinger equation in exactly the same form as the original Eq. (7-1) with the understanding that it relates to electronic motion only.

There is a very convenient way of writing the hamiltonian for atomic and molecular systems once the Born–Oppenheimer approximation has been made. One simply writes a kinetic energy part $-\frac{1}{2}\nabla^2$ for each nonstationary particle and a coulombic potential $\pm Z/r$ for each interparticle electrostatic interaction. The form of the potential is such that it has $\pm Z$ in the numerator, where Z is the charge, and the interparticle distance r in the denominator. The sign is $+$ for repulsion and $-$ for attraction. Thus the hamiltonian for the hydrogen atom has $-1/2\nabla^2$ kinetic energy of its single moving electron and $-1/r$ as the electrostatic potential energy, where $Z = 1$, the force is attractive, and there is one electron at a distance r:

$$H = -\frac{1}{2}\nabla^2 - \frac{1}{r}$$

EXERCISE 7-1

Write the hamiltonian for the helium atom, which has two electrons, one at a distance r_1 and the other at a distance r_2.

Solution 7-1

$$H = -\frac{1}{2}\nabla_1^2 - \frac{1}{2}\nabla_2^2 - \frac{2}{r_1} - \frac{2}{r_2} + \frac{1}{r_{12}}$$

Notice the interelectronic repulsion term $+1/r_{12}$.

These equations seem to be leaving a lot of information out, but they can be shown to be mathematically identical to longer forms, depending on the arbitrary choice of units. The short forms, which we shall use extensively, are scaled by appropriate unit choices so that several constants become one and much notation cancels out. The "natural" system of units for these problems is called the system of *atomic units* (a.u.; it is discussed by Mc Quarrie (Mc Quarrie, 1983).

Because the lowest energy solution of the Schroedinger equation for hydrogen is ψ_{1s}, this is often given the short symbol $1s$. Using short notation for hydrogen, the Schroedinger equation for the ground state can be written

$$\left(-\frac{1}{2}\nabla^2 - \frac{1}{r}\right)1s = E(1s) \tag{7-2}$$

By extension of Exercise 7-1, the hamiltonian for a many-electron molecule (the kind we are interested in) should have a sum of kinetic energy operators $-\frac{1}{2}\nabla^2$, one for each electron. Also, each electron moves in the potential of the nuclei and all other electrons; therefore,

$$H = \sum -\tfrac{1}{2}\nabla_i^2 + \sum V_i$$

where the terms V_i may be $+$ or $-$ according to whether the electron is attracted (to nuclei) or repelled (by other electrons). One can split up the attractive and repulsive terms:

$$H = \sum\left(-\tfrac{1}{2}\nabla^2(i) - V(i)\right) + \tfrac{1}{2}\sum 1/r_{ij} \tag{7-3}$$

where the repulsive term is multiplied by $\frac{1}{2}$ because we do not want to count repulsions twice, e.g., as *i-j* repulsions and again as *j-i* repulsions.

This hamiltonian operator is called an *n*-electron hamiltonian,

$$H = H(1, 2, 3, \ldots, n)$$

The symbol $-\frac{1}{2}\nabla^2(i)$ is the kinetic energy operator for the *i*th electron (Mc Quarrie, 1983) and $V(i)$ is the potential energy of the *i*th electron in the field of all nuclei but in the absence of all other electrons.

The reason the Schroedinger equation for molecules cannot be separated appears in the last term $\frac{1}{2}\Sigma 1/r_{ij}$, involving a sum of repulsive energies between electrons. In order to obtain this sum, or even one term of it, say for the ith electron, one must know exactly where the jth electron is. This is because repulsive force is dependent on distance. The position of the jth electron depends on the position of the ith electron, however, which is what we are trying to find.

Pi-Electron Separation Approximation

We assume that the chemistry of unsaturated hydrocarbons is so dominated by the chemistry of their double bonds that we may separate the Schroedinger equation into an equation for sigma electrons and one for pi electrons. We ignore the sigma electrons, as we did the nuclei, except for the contribution to the potential energy brought about by their charge. We now have an equation of the same form as Eq. (7-3), but one in which the hamiltonian for all electrons is replaced by the hamiltonian for pi electrons only,

$$H = \sum^{n} \left(-\tfrac{1}{2}\nabla^2(i) - V_{pi}(i)\right) + \tfrac{1}{2}\sum^{n} 1/r_{ij} \qquad (7\text{-}4)$$

with somewhat different meanings for the symbols. Now, n is the number of pi electrons of kinetic energy $-\frac{1}{2}\nabla^2$ and the potential energy term V_{pi} represents the potential energy of a single pi electron in the field of the framework of nuclei and sigma electrons.

Single-Electron Approximation

The r_{ij} term is still the problem. The simplifying assumption is made that $V'(i)$ is an average potential due to the nuclei and the electrons other than electron i:

$$H(i) = -\tfrac{1}{2}\nabla^2(i) - V'(i) \qquad (7\text{-}5)$$

With this new approximation, the r_{ij} term does not appear [it is hidden in $V'(i)$] and the Schroedinger equation becomes separable into n equations, all the same, one for each pi electron:

$$H\psi_i = E_i\psi_i \qquad (7\text{-}6)$$

where H includes the new potential energy term V'. This term is unknown (the r_{ij} problem has not gone away); hence, no closed solution exists for Eq. (7-6).

To obtain an approximate solution of Eq. (7-6), we select any set of basis functions,

$$\psi = \sum a_i\phi_i \qquad (7\text{-}7)$$

as an approximation to the true wave function. If the basis set ϕ is judiciously chosen, ψ may be a good approximation to the true ψ. We shall now look at one way of choosing a basis set and develop an iterative procedure to obtain the best ψ from any $\Sigma a_i \phi$.

THE HUCKEL METHOD

Although in the Huckel approximation, pi electrons are not supposed to interact with sigma electrons, they do exchange with one another. In the 1920s, Heitler and London showed that the chemical bond existing between two identical hydrogen atoms in H_2 can be described mathematically by taking a linear combination of the $1s$ orbitals of the two H atoms that are partners in the molecule. When this is done, the combination

$$\Psi = a_1 \psi_{1s_1} + a_2 \psi_{1s_2} \qquad (7\text{-}8)$$

is a new solution of the Schroedinger equation that has the characteristics of the chemical bond; specifically, a unique low-energy internuclear distance that approximates the bond distance in H_2. The right side of Eq. (7-8) is called a *linear combination of atomic orbitals* (LCAO). The Heitler–London method is also known as the *valence bond approximation* (VB). It arrives at a description of chemical bonding from a somewhat different logical premise than that of the *molecular orbital* (MO) method. Both methods arrive at equivalent mathematical descriptions, as they must, because they are attempts to describe the same thing.

The LCAO method is only one of infinitely many ways that a molecular orbital can be approximated. At this point, we shall restrict ourselves to a purely LCAO–MO description of organic molecules, although many other applications of both VB and MO methods exist.

In 1930, Huckel showed that the LCAO approximation can be applied to the single electrons of the p orbitals of carbon atoms that are partners in a $C{=}C$ bond. The p orbitals are, of course, considered to be independent of the sigma-bonded framework except for the potential energy of charge interaction [Eq. (7-6)]. The linear combination is

$$\psi = \sum a_i p_i \qquad (7\text{-}9)$$

where the p_i are the p orbitals of double-bonded carbon atoms (Fig. 7-2). When this is done for ethylene, for example,

$$\psi_1 = a_1 p_1 + a_2 p_2 \qquad (7\text{-}10a)$$

$$\psi_2 = a_1 p_1 - a_2 p_2 \qquad (7\text{-}10b)$$

FIGURE 7-2 The pi orbitals of ethylene.

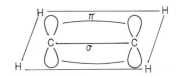

It is a property of linear, homogeneous differential equations, of which the Schroedinger equation is one example, that a solution multiplied by a constant is also a solution and a solution added to a solution is also a solution. Therefore, if the solutions p_1 and p_2 in Eqs. (7-10) were exact molecular orbitals, ψ would also be exact. Orbitals p_1 and p_2 are not exact MOs; they are exact orbitals of the hydrogen atom.

The Expectation Value of the Energy

Premultiplying each side of the Schroedinger equation by ψ, gives

$$\psi E\psi = \psi H\psi$$

but E is a scalar; hence,

$$E\psi\psi = \psi H\psi$$

for one electronic configuration in ethylene. For all electronic configurations, one must integrate over all space $d\tau$

$$E\int \psi^2\, d\tau = \int \psi H\psi\, d\tau$$

or

$$E = \frac{\int \psi H\psi\, d\tau}{\int \psi^2\, d\tau} \tag{7-11}$$

It is a fundamental *postulate* of quantum mechanics that E is the expectation value of the energy for wave function ψ. If the values of ψ are exact, E is exact. If the ψs are approximate, as they are in this case, E is an upper bound to the true energy. Minimizing E will be used to obtain the best value of ψ from a given basis set. Other criteria may be selected, leading to different estimates of how closely ψ approximates its exact value. All properties of the system approach their true values as ψ approaches its exact value.

Substituting the first LCAO for ψ, we have

$$E = \frac{\int (a_1 p_1 + a_2 p_2) H (a_1 p_1 + a_2 p_2)\, d\tau}{\int (a_1 p_1 + a_2 p_2)^2\, d\tau}$$

Expanding this equation yields four integrals in the numerator and four in the denominator. This takes a lot of space, so we use the notation

$$\int p_1 H p_1\, d\tau = \alpha$$

$$\int p_1 H p_2\, d\tau = \beta$$

$$\int p_1 p_1\, d\tau = S_{11} \tag{7-12}$$

$$\int p_1 p_2\, d\tau = S_{12}$$

We also assume that the order of the subscripts on the atomic orbitals p is immaterial in writing α, β, and S. Some of these assumptions are not self-evident. The interested reader should consult a quantum mechanics text (e.g., Hanna, 1981) for their justification or critique.

The expression for the energy, after all assumptions and notation simplifications have been made, is

$$E = \frac{a_1^2\alpha + 2a_1a_2\beta + a_2^2\alpha}{a_1^2S_{11} + 2a_1a_2S_{12} + a_2^2S_{22}} \tag{7-13}$$

If we could evaluate α, β, and S, which are called the coulomb, exchange, and overlap integrals respectively, we could compute E.

We still do not know either side of the equation, but we do know that E is to be minimized with respect to some minimization parameters. The only arbitrary parameters we have are the a_{ij} that enter into the LCAO. Thus our minimization criteria are

$$\frac{\partial E}{\partial a_1} = 0 \tag{7-14a}$$

$$\frac{\partial E}{\partial a_2} = 0 \tag{7-14b}$$

These minimizations lead to

$$a_1\alpha + a_2\beta = E(a_1S_{11} + a_2S_{12}) \tag{7-15a}$$

$$a_1\beta + a_2\alpha = E(a_1S_{12} + a_2S_{22}) \tag{7-15b}$$

or

$$a_1(\alpha - ES_{11}) + a_2(\beta - ES_{12}) = 0 \tag{7-16a}$$

$$a_1(\beta - ES_{12}) + a_2(\alpha - ES_{22}) = 0 \tag{7-16b}$$

These are the normal equations for the minimization.

A further simplification is made. The wave functions p_1 and p_2, which are orthonormal in the hydrogen atom, are assumed to retain their orthonormality in the molecule. Orthonormality requires that

$$S_{11} = S_{22} = \int p_1p_2 \, d\tau = \int p_2p_2 \, d\tau = 1$$

and

$$S_{12} = S_{21} = \int p_1p_2 \, d\tau = \int p_2p_1 \, d\tau = 0$$

This yields

$$a_1(\alpha - E) + a_2\beta = 0 \tag{7-17a}$$

$$a_1\beta + a_2(\alpha - E) = 0 \tag{7-17b}$$

as the normal equations. In this context, the normal equations are also called *secular equations*.

Homogeneous Simultaneous Equations

What we formerly called the *nonhomogeneous vector* (Chapter 4) is zero in Eqs. (7-17). When this vector vanishes, the set is homogeneous. Let us try to construct a simple set of linearly independent homogeneous equations:

$$x + y = 0$$
$$x + 2y = 0$$

The set cannot be true for any solution set other than $\{0, 0\}$. Again,

$$x + y = 0$$
$$3x + y = 0$$

cannot be true except for $\{x, y\} = \{0, 0\}$.

Any linearly independent set of homogeneous equations we can construct has only the zero vector as its solution set. This is not acceptable because it means that the wave function vanishes, which is contrary to hypothesis (the electron has to be somewhere). We are driven to the conclusion that the normal equations must be linearly dependent.

Linearly dependent sets of homogeneous simultaneous equations, e.g.,

$$x + 2y = 0$$
$$2x + 4y = 0 \tag{7-18a}$$

have infinitely many solution sets, e.g., $\{2, -1\}$, $\{2.2, -1.1\}$, $\{2.4, -1.2\}$, etc. In order to select one from among the infinite number of solution sets, one must have an additional independent nonhomogeneous equation. If the additional equation is

$$x + y = 1 \tag{7-18b}$$

the solution set $\{2, -1\}$ satisfies all three equations and is the unique solution set for the homogeneous linear simultaneous equation plus the additional equation.

In what follows, we shall obtain eigenvalues (Chapter 3) for the simultaneous solution set (7-17). Each eigenvalue gives a pi-electron energy for the model we used to generate the secular equation set. In the next chapter, we shall apply an additional equation of constraint on the minimization parameters $\{a_1, a_2\}$ so as to obtain their unique solution set.

The Secular Matrix

The coefficient matrix of the normal equations for ethylene is

$$\begin{pmatrix} \alpha - E & \beta \\ \beta & \alpha - E \end{pmatrix} \tag{7-19}$$

It should be evident that E is an eigenvalue of the matrix in α and β (Example 3-11). There are two secular equations in two unknowns for ethylene. For a system with n conjugated sp^2 carbon atoms, there will be n secular equations leading to n eigenvalues E. Each secular equation yields a new eigenvalue and a new eigenvector (see Chapter 8).

If we divide each element of the secular matrix by β and perform the substitution $x = (\alpha - E)/\beta$, we get

$$\begin{pmatrix} x & 1 \\ 1 & x \end{pmatrix} \tag{7-20}$$

Diagonalizing the Ethylene Coefficient Matrix

The secular equations [Eqs. (7-17)] can be written in matrix form:

$$\begin{pmatrix} x & 1 \\ 1 & x \end{pmatrix}\begin{pmatrix} a_1 \\ a_2 \end{pmatrix} = 0$$

We have already said that the solution vector cannot be zero, $\{a_1, a_2\} \neq \{0, 0\}$; hence, the matrix of coefficients must be zero:

$$\begin{pmatrix} x & 1 \\ 1 & x \end{pmatrix} = 0$$

This matrix can be written as a sum

$$\begin{pmatrix} x & 0 \\ 0 & x \end{pmatrix} + \begin{pmatrix} 0 & 1 \\ 1 & 0 \end{pmatrix} = 0$$

or $\tag{7-21}$

$$\begin{pmatrix} x & 0 \\ 0 & x \end{pmatrix} = -\begin{pmatrix} 0 & 1 \\ 1 & 0 \end{pmatrix}$$

If we can diagonalize the matrix on the right of the equal sign, we shall have obtained the solution set for x. By substituting back into the definition of x, we shall have the solution set for E, the energy.

Rotation Matrices

If we premultiply and postmultiply the matrix

$$\begin{pmatrix} 0 & 1 \\ 1 & 0 \end{pmatrix}$$

by the matrix

$$\begin{pmatrix} \cos\theta & \sin\theta \\ \sin\theta & -\cos\theta \end{pmatrix}$$

where $\theta = 45°$, the result is

$$\begin{pmatrix} 1 & 0 \\ 0 & -1 \end{pmatrix}$$

which is the original matrix rotated one-eighth turn or 45° with a sign change. The premultiplying and postmultiplying matrix is often called a rotation matrix. The matrix

$$\begin{pmatrix} \cos\theta & -\sin\theta \\ \sin\theta & \cos\theta \end{pmatrix}$$

is widely used to do the same thing.

This is the result we were looking for. By rotating it through the proper angle, we have diagonalized the matrix on the right of (7-21):

$$\begin{pmatrix} x & 0 \\ 0 & x \end{pmatrix} = -\begin{pmatrix} 1 & 0 \\ 0 & -1 \end{pmatrix} = \begin{pmatrix} -1 & 0 \\ 0 & 1 \end{pmatrix}$$

Diagonalization yields the solution set $x = \{-1, 1\}$. Multiplying by B,

$$\alpha - E = \mp B$$

or   (7-22)

$$E = \alpha \mp B$$

We can represent the energy of the system by a diagram with the atomic orbital energy at α and the molecular orbital energies (there are two) at β above and β below the AOs. This orbital diagram should be familiar from elementary treatments of homonuclear diatomic molecules like H_2. In this treatment, the two molecular orbitals are the bonding and antibonding pi orbitals of the ethylene molecule. The bonding orbital is the lower one and the antibonding orbital is the upper one.

Generalization

The advantage of the method just described is that it can be generalized to molecules of any size. Setting up quite complicated secular matrices can be reduced to a simple recipe. A computer scheme can be used to diagonalize the resulting matrices by an iterative series of rotations.

The dimension of the matrix is the number of atoms in the pi conjugated system. Let us take the three-carbon system allyl as our next step. Concentrate on one atom in the system and write an x in the 1, 1 position for that atom. Follow this by writing a 1 in the position corresponding to any atom attached to the x atom. For any atom not attached to x, enter a zero. For allyl

$$\begin{array}{c} C-C-C \\ \uparrow \end{array}$$

this leads to the top row of the secular matrix

$$x \quad 1 \quad 0$$

Concentrating on the second atom in the allyl chain leads to the row

1 x 1 and concentrating on the third atom gives 0 1 x. The full allyl matrix is

$$\begin{pmatrix} x & 1 & 0 \\ 1 & x & 1 \\ 0 & 1 & x \end{pmatrix}$$

The zeros in the 1, 3 and 3, 1 positions correspond physically to assuming that there is no interaction between p electrons of atoms that are not neighbors. This is a standard assumption of Huckel theory.

If we had been interested in the cyclopropenyl system, we would have been led to the matrix

$$\begin{pmatrix} x & 1 & 1 \\ 1 & x & 1 \\ 1 & 1 & x \end{pmatrix}$$

Butadiene yields

$$\begin{pmatrix} x & 1 & 0 & 0 \\ 1 & x & 1 & 0 \\ 0 & 1 & x & 1 \\ 0 & 0 & 1 & x \end{pmatrix}$$

and so on.

The rotation matrix must also be given in general form. If the premultiplying and postmultiplying matrix is contained as a block within a larger matrix containing only 1s on the principal diagonal and 0s elsewhere (aside from the rotation block), only the corresponding block of the operand matrix (the matrix that is operated on) is rotated. Elements outside the rotation block are changed too. For example,

$$\begin{pmatrix} \cos\theta & \sin\theta & 0 \\ \sin\theta & -\cos\theta & 0 \\ 0 & 0 & 1 \end{pmatrix} \begin{pmatrix} 1 & 1 & 0 \\ 1 & 0 & 1 \\ 0 & 1 & 0 \end{pmatrix} \begin{pmatrix} \cos\theta & \sin\theta & 0 \\ \sin\theta & -\cos\theta & 0 \\ 0 & 0 & 1 \end{pmatrix}$$
$$= \begin{pmatrix} 1 & 0 & 0.707 \\ 0 & -1 & -0.707 \\ 0.707 & -0.707 & 0 \end{pmatrix}$$
(7-23)

where θ was once again taken as 45°. The rotation matrix R can be made as large as necessary to conform with any operand matrix. The rotation block may be placed anywhere on the principal diagonal of the rotation matrix.

The Jacobi Method

The Jacobi method is probably the simplest diagonalization method that is well adapted to computers. It is limited to real symmetric matrices, but that is the only kind we will get by the formula for generating simple Huckel molecular orbital (HMO) matrices just described. A rotation matrix is defined as

$$\mathbf{R} = \begin{pmatrix} 1 & 0 & 0 & 0 & \text{etc} \\ 0 & \cos\theta & \sin\theta & 0 & \\ 0 & \sin\theta & -\cos\theta & 0 & \\ 0 & 0 & 0 & 1 & \\ \text{etc.} & & & 1 & \\ & & & & 1 & \\ & & & & & \text{etc.} \end{pmatrix} \qquad (7\text{-}24)$$

so as to "attack" a block of elements of the operand matrix. In the case of rotation matrix (7-24), the block with a_{22} and a_{33} on the principal diagonal is attacked. Now,

$$\tan 2\theta = \frac{2a_{ij}}{a_{ii} - a_{jj}} \qquad (7\text{-}25)$$

where a_{ij} denotes the ij element in matrix $\mathbf{A}$ and a_{ii} and a_{jj} are on the principal diagonal. The matrix equation

$$\mathbf{RAR} = \mathbf{A'} \qquad (7\text{-}26)$$

generates a matrix $\mathbf{A'}$, which is *similar* to $\mathbf{A}$ but has had elements a_{ij} and a_{ji} reduced to 0. The good news is that any a_{ij} and a_{ji} elements not on the principal diagonal can be converted to 0 by choosing the right $\mathbf{R}$ matrix. The bad news is that each successive $\mathbf{RAR}$ multiplication destroys all 0s previously gained, replacing them with elements that are not 0 but are smaller than their previous value. Thus, the $\mathbf{RAR}$ multiplication must be carried out a number of times that is not just equal to one-half the number of nonzero off-diagonal elements, but that is very large, strictly speaking, infinite. The sum of the off-diagonal elements cannot be set equal to 0 by the Jacobi method, but it can be made to converge on 0; the Jacobi method is an iterative method.

Let us follow the first few iterations for the ALLYL system by hand calculations. We subtract the matrix $x\mathbf{I}$ from the HMO matrix to obtain the matrix we wish to diagonalize, just as we did with ethylene. With the rotation block in the upper left corner of the $\mathbf{R}$ matrix (we are attacking a_{12} and a_{21}), we wish to find

$$\mathbf{R} \begin{pmatrix} 0 & 1 & 0 \\ 1 & 0 & 1 \\ 0 & 1 & 0 \end{pmatrix} \mathbf{R}$$

First, we find that, concentrating on $a_{ij} = a_{12}$,

$$\tan 2\theta = \frac{2a_{ij}}{a_{ii} - a_{jj}} = \infty$$

$$\theta = 45°$$

$$\sin\theta = \cos\theta = \frac{1}{(2)^{1/2}} = 0.7071$$

By the simple HMO procedure, it is always true that $\sin \theta = \cos \theta = 0.7071$ on the first iteration. Now, to eliminate a_{12} and a_{21},

$$\mathbf{RAR} = \mathbf{A}' \qquad (7\text{-}27)$$

but this is just the multiplication we used as an illustration in the last section. We know that the result is matrix (7-23).

Elements a_{12} and $a_{21} = 0$. We have gained the 0s we wanted, but we have sacrificed the 0s we had in the 1, 3 and 3, 1 positions. Other than those eliminated, the off-diagonal elements are no longer 0, but they are less than 1. Attacking the $a_{13} = a_{31} = 0.7071$ elements produces

$$A'' = \begin{pmatrix} 1.37 & -0.325 & 0 \\ -0.325 & -1 & 0.628 \\ 0 & 0.628 & -0.37 \end{pmatrix}$$

Nine iterations yield

$$A''''''' = \begin{pmatrix} -1.41 & 0 & 0 \\ 0 & 0 & 0 \\ 0 & 0 & 1.41 \end{pmatrix}$$

and the energy levels or eigenvalues for the three-carbon allyl model

$$x = -1.41, 0, 1.41$$

which are now on the principal diagonal.

The order of the roots as generated by diagonalization is not, in general, the order we have given, lowest to highest. It is dependent upon the algorithm, as are some of the intermediate matrices generated in the diagonalization procedure. Programs are written to be "opportunistic," that is, to seek a quick means of conversion on the eigenvalues; the strategy chosen may differ from one program to the next. Many programs, including the one to be described in the next chapter have a separate subroutine that takes the eigenvalues in whatever order they are produced by diagonalization and orders them, lowest to highest or vice versa.

PROGRAM MOBAS

A simple method of generating the eigenvalues for the general HMO matrix (Dickson, 1968; Rogers et al., 1983) involves searching the HMO matrix to find the largest off-diagonal element, i.e., the one most suitable for attack. Once this element is found, the rotation angle is calculated by Eq. (7-25) and the matrix is partially diagonalized. The search is repeated, the next target element is selected, the rotational angle is calculated, and so on. After each rotation, the off-diagonal elements are tested to see whether they are sufficiently close to 0. Of course they are not at the beginning, but they get smaller as the rotation is iterated. An arbitrary standard is set up so that when the diagonal elements have been reduced below a certain level, the rotation iterations stop. In MOBAS, the criterion for exit from the iterative diagonalization loop is that the root mean square sum of the off-diagonal elements be equal to or less than 10^{-7} times its original value.

Degenerate roots (different roots having the same energy) can produce computational difficulties. These problems can usually be circumvented by entering the HMO matrix with elements that are slightly different from 1. For example, 1.0001 might be used.

Heterocyclic and linear heteronuclear pi conjugated systems pose a special problem because the heteroatom has an electron density that is greater than or less than the electron density of carbon. Let us take the nitrogen in pyrrole, which is electron rich, as an example.

The Jacobi procedure suggests an empirical method of compensating for the increased electron density at nitrogen in the way in which elements on the principal diagonal are built up by accretion during the iterative diagonalization procedure. If we place a nonzero element on the principal diagonal of the matrix to be diagonalized (after the x matrix has been subtracted), when the accretion process is over, that position will have an energy lower (or higher depending on the sign of the root) than it otherwise would have had.

In pyrrole, the value of 1.5 in the 1, 1 position causes the lowest energy level to be lowered and the electron density about the nitrogen to be larger than it would be for carbon, which has a 0 in the 1, 1 position. The value 1.5 is selected by trial and error by comparison to experimental values for spectral transitions, resonance energies, etc., and represents a literature consensus (Strietwieser, 1961). Empirical modification of off-diagonal entries in the HMO matrix are also used for bonds connecting carbon to other atoms.

For pyrrole with 1.5 in the lead position of the HMO matrix, 31 iterations yield a lowest eigenvalue of -2.55β: $E = \{-2.55, 1.20, -1.15, 1.62, 0.62\}$. This will be taken up in more detail in the next chapter.

COMPUTER PROJECT 7-1 | *Energy-Level Manifolds*

An energy-level manifold is a collection of quantum-mechanical energy levels arranged in increasing order from bottom to top. Each energy level coincides with an eigenvalue of the Schroedinger equation. In Huckel molecular orbital theory, energies are given in units of β relative to α, which is arbitrarily taken to be 0. Manifolds are often presented as diagrams like Fig. 7-3. The wave function for the higher of the two energies in Fig. 7-3 has one internal node, but the lower energy function has no internal nodes. This is general: The more internal nodes, the higher

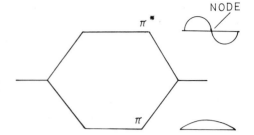

FIGURE 7-3 The energy-level manifold and nodal properties of the molecular orbitals of ethylene.

the energy. Energy-level manifolds are usually more complicated than Fig. 7-3 and have many levels for large molecules.

Illustration: The Allyl Model The Huckel matrix for the allyl model has already been given. Its solution yields three eigenvalues and three eigenfunctions, with zero, one, and two internal nodes. Draw the energy-level manifold for the allyl model. Any energy below α in energy is bonding; label it π. Any level above α is antibonding; label it π^*. One can order the eigenvalues produced by observing the number of times the eigenfunction changes sign.

The number of sign changes (nodes) can be determined from the sign of the coefficients associated with the eigenvalue. For example, the eigenvalue 1.414 is associated with coefficients having $-+-$ or $+-+$ signs. There are two changes in sign; hence, there are two internal nodes. The term *internal node* means that we ignore the nodes at either end of the wave function, which are the same for all wave functions and provide no useful information.

Procedure Execute MOBAS and determine the energy levels (eigenvalues) for the ethene, allyl, butadiene, and pentadiene models. Draw diagrams like Fig. 7-3 that show the energy levels in their proper order, lowest to highest. If an energy turns out to be 0 (relative to α), label it n: nonbonding. Remember that, because of rounding and a finite number of matrix rotations, the zero roots may appear as very small values, say 10^{-7} or so:

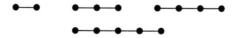

Using MOBAS, calculate and order the eigenvalues for the cyclopropene, cyclobutadiene, and cyclopentadiene models. Draw the energy-level manifolds and compare them with the preceding linear models. Two roots with the same energy are said to be degenerate. They are not duplicate solutions to the Schroedinger equation because they have different coefficients (eigenfunctions). See Chapter 8 for a discussion of eigenfunctions. Are there any degenerate roots among these model systems?

COMPUTER PROJECT 7-2 | HMO *Calculations of Spectroscopic Transitions*

Linear polyenes (butadiene, hexatriene, etc.) absorb ultraviolet radiation. Their absorption maxima occur at the wavelengths given in Table 7-1.

These absorptions are ascribed to π-π^* transitions, that is, transitions of an electron from the highest occupied pi orbital to the lowest unoccupied pi orbital. Because these are molecular orbitals, they are called the HOMO and LUMO, respectively. One can decide which orbitals are the HOMO and LUMO by filling electrons into the molecular energy-level manifold from the bottom up. The number of electrons is the number of sp^2 carbon atoms in a neutral polyalkene. In ethylene, there is only one occupied MO and one unoccupied MO. The occupied orbital in ethylene is β below the energy level represented by α and the unoccupied orbital is β above it. The separation of the HOMO and LUMO is 2β.

Using MOBAS and the method from Computer Project 7-1, calculate the energy separation between the HOMO and LUMO in units of β for all compounds in Table 7-1 and enter the results in Table 7-2. Enter the observed energy of ultraviolet radiation absorbed for each compound in units of cm^{-1}. The reciprocal wavelength is often used as a unit of

TABLE 7-1

Ethylene	161 nm
Buta-1,3-diene	217 nm
Hexa-1,3,5-triene	244 nm
Octa-1,3,5,7-tetraene	303 nm

TABLE 7-2

Compound	HOMO	LUMO	(LUMO − HOMO)	$1/\lambda$, cm^{-1}
Ethene	$\alpha + \beta$	$\alpha - \beta$	-2β	$\cdots$
Buta-1,3-diene	$\cdots$			
$\cdots$				
etc.				

spectroscopic energy. Radiation of wavelength 161 nm, for example, has an energy of 6.21×10^4 cm^{-1}.

The quantity β is an energy. The separation in MO levels (2β in the case of ethylene) is a change in energy E and follows Planck's equation $\Delta E = h\nu$. You should be able to calculate four energies of radiation absorbed to promote an electron across four different energy gaps, measured in units of β. Using program LLSQ, obtain the best slope of the function E cm^{-1} vs. β. This is the amount of energy in cm^{-1} per β, i.e., the "size" of the energy unit β. The calculation is approximate, but even an order of magnitude calculation of β is useful. Calculate β in units of electron volts (eV); we will need it later.

PROBLEMS|*Chapter 7*

1 Show that Eq. (7-15) follows from Eq. (7-14).

2 Compute the HMO eigenvalues for bicyclobutadiene.

3 Draw the energy-level diagram for bicyclobutadiene.

4 Write the secular matrix for methylenecyclopropene.

5 Compute the eigenvalues and eigenvectors for methylenecyclopropene.

6 Draw the energy-level diagram for methylenecyclopropene.

7 Write the secular matrix for fulvene.

8 Compute the eigenvalues and eigenvectors for fulvene.

9 Draw the energy-level diagram for fulvene.

10 Compute the HMO eigenvalues for benzene and draw its energy-level diagram.

11 Compute the HMO eigenvalues for cyclopropenone. Obtain the parameter for O from the literature.

12 Draw the energy-level diagram for pyrrole.

Place a 2 on the principal diagonal for N. Make no alteration in β for the C — N bond. Compare with the value on p. 95, which was obtained using 1.5 on the principal diagonal for N.

REFERENCES

Atkins, P. W., 1986. *Physical Chemistry*. W. H. Freeman, New York.

Dickson, T. R., 1968. *The Computer and Chemistry*. Freeman, San Francisco, CA.

Dirac, P. A. M., 1929. *Proc. Roy. Soc. London*, **A123**, 714.

Hanna, M. W., 1981. *Quantum Mechanics in Chemistry*, 3rd ed. Benjamin, Menlo Park, CA.

Mc Quarrie, D. A., 1983. *Quantum Chemistry*. University Science Books, Mill Valley, CA.

Rogers, D. W., Angelis, B. P., and Mc Lafferty, F. J., 1983. *Amer. Lab.* **15** (11), 54.

Streitwieser, A., 1961. *Molecular Orbital Theory for Chemists*. Wiley, New York.

8 Molecular Orbital Calculations II

This chapter continues the discussion of molecular orbital calculations begun in Chapter 7 with special attention to the eigenfunctions.

THE MATRIX AS OPERATOR

An operator is a mathematical instruction; for example, the operator d/dx is the instruction to differentiate once with respect to x. Matrices in general, and the R matrix of Chapter 7 in particular, are operators. The R matrix is an instruction to rotate a part of the operand matrix **A** through a certain angle.

The product of matrix operators is an operator. Rotation through 90°, for example, followed by another rotation in the same direction and in the same plane through 90° is the same as one rotation through 180°:

$$\mathbf{R}_{90}\mathbf{R}_{90} = \mathbf{R}_{180}$$

Thus, the Jacobi procedure, by making many rotations of the elements of the operand matrix, ultimately arrives at the operator matrix that diagonalizes it. Mathematically, we can imagine one operator matrix that would have diagonalized the operand matrix all in one step

$$\mathbf{R}_t = \mathbf{R}_1\mathbf{R}_2\mathbf{R}_3 \cdots \mathbf{R}_n$$

even though in practice, we took n steps to do it. The situation is analogous to tuning a radio. We can imagine one perfect twist of the dial that would land right on the station, but in practice, we make several little twists back and forth across the proper tuning until we find the tuning we like.

COMPUTATION OF THE COEFFICIENT MATRIX

The general form of the secular matrix equation for the Huckel molecular orbital theory is

$$(\mathbf{H} - E\mathbf{I})\mathbf{a} = 0 \tag{8-1}$$

where **a** is a solution vector with coefficients $\{a_j\}$. There are n of them, one for each of the eigenvalues. The H matrix consists of elements (operators)

$$H_{ii} = \alpha_i = \int \phi_i H \phi_i \, d\tau$$

and

$$H_{ij} = \beta_{ij} = \int \phi_i H \phi_j \, d\tau$$

leading to

$$\begin{pmatrix} \alpha_1 - E & \beta_{12} & 0 & \cdots & 0 \\ \beta_{21} & \alpha_2 - E & \beta_{23} & \cdots & 0 \\ & \cdots & & & \\ 0 & 0 & \cdots & \beta_{nn-1} & \alpha_n - E \end{pmatrix} \begin{pmatrix} a_1 \\ a_2 \\ \cdots \\ a_n \end{pmatrix} = 0 \quad (8\text{-}2)$$

where orbital interactions that do not involve nearest-neighbor atoms are taken to be zero. For example, $\beta_{13} = 0$. The vector **a** is the set of coefficients for one eigenfunction ψ_i corresponding to one eigenvalue. From Eq. (8-1),

$$\begin{aligned} \mathbf{Ha} - E_i \mathbf{Ia} &= 0 \\ \mathbf{Ha} &= E_i \mathbf{Ia} = \mathbf{a} E_i \end{aligned} \quad (8\text{-}3)$$

If the eigenvector is found for each of the n eigenvalues E_i and arranged column by column in the same order as the eigenvalues with which they are associated, they form an $n \times n$ matrix **A** with elements that are all the coefficients of all the basis functions for all the eigenvalues. The single coefficient vector is replaced by the entire matrix of coefficient vectors.

Now

$$\mathbf{HA} = \mathbf{AE} \quad (8\text{-}4)$$

where **E** is a diagonal matrix with the eigenvalues along the diagonal. We can multiply both sides by $\mathbf{A}^{-1}$:

$$\mathbf{A}^{-1}\mathbf{HA} = \mathbf{A}^{-1}\mathbf{AE} = \mathbf{E} \quad (8\text{-}5)$$

but **A** is an orthonormal matrix and has the property that it is its own inverse. Hence,

$$\mathbf{AHA} = E_i \mathbf{I} = \mathbf{E} \quad (8\text{-}6)$$

but this is the same mathematical operation that we used to obtain the diagonalized matrix of the eigenvalues, except that **H** was pre- and postmultiplied by the rotation matrix. The matrix of the coefficients (which we have called **A** in this section) must be the total rotation matrix (which we called **R** in Chapter 6 or R_t in the first section of this chapter):

$$A = R_t = AI = R_t I \tag{8-7}$$

Thus, if, when we are finding the total rotation matrix as a product of partial rotation matrices, we also allow that series of partial rotation matrices to operate on the unity matrix, we shall generate the matrix of the coefficients **A**.

This is exactly what is done in programs MOBAS and SHMO (Simple Huckel Molecular Orbital). The secular matrix is diagonalized by repeated rotation

$$RXR \Rightarrow EI \tag{8-8}$$

until the correct rotation matrix R_t has been found. The matrix of coefficients of the wave functions is produced by performing the same operations on the unit matrix

$$R_1 R_2 R_3 \cdots R_n I = R_t I = A$$

where

$$A\{p_i\} = \{\psi_i\}$$

for a basis set of p orbitals. For example, for ethylene,

$$a_{11} p_1 + a_{12} p_2 = \psi_1$$
$$a_{21} p_1 + a_{22} p_2 = \psi_2 \tag{8-9}$$

Let us try a verification for ethylene. The rotation matrix is

$$\begin{pmatrix} \cos \theta & \sin \theta \\ \sin \theta & -\cos \theta \end{pmatrix} = \begin{pmatrix} 0.707 & 0.707 \\ 0.707 & -0.707 \end{pmatrix}$$

for $\theta = 45°$.

Allowing this matrix to operate on the identity matrix **I** simply gives the rotation matrix **R**. We are led to believe that the elements of the rotation matrix (for $\theta = 45°$) are the coefficients of the wave functions

$$0.707 p_1 + 0.707 p_2 = \psi_1$$
$$0.707 p_1 - 0.707 p_2 = \psi_2 \tag{8-10}$$

To check this, substitute the eigenvalues $x = \{1, -1\}$ into the secular equations

$$a_1 x + a_2 = 0$$
$$a_1 = -a_2$$

for the first root and

$$a_1 + a_2 x = 0$$
$$a_1 = a_2$$

for the second.

We have only a ratio $a_1 = \pm a_2$ because the secular equations are not linearly independent. If we impose the normalization condition

$$a_1^2 + a_2^2 = 1$$

then, because of symmetry, $a_1 = a_2$, hence

$$2a_1^2 = 1$$

$$a_1 = \pm(2)^{-1/2} = 0.707$$

that is,

$$0.707p_1 + 0.707p_2 = \psi_1$$
$$0.707p_1 - 0.707p_2 = \psi_2$$

(8-11)

which is the same result that we previously obtained.

EXERCISE 8-1

Execute MOBAS (on disk) using the allyl model

$$\begin{pmatrix} 0 & 1 & 0 \\ 1 & 0 & 1 \\ 0 & 1 & 0 \end{pmatrix}$$

Arrange the coefficients of the eigenfunctions as the columns of a 3×3 matrix with the low-energy eigenvector on the left and the high-energy eigenvector on the right. Each column is an eigenvector $\mathbf{a}$. The entire matrix is the eigenvector matrix $\mathbf{A}$.

Solution 8-1

$$\mathbf{A} = \begin{pmatrix} 0.5 & 0.707 & 0.5 \\ 0.707 & 0 & -0.707 \\ 0.5 & -0.707 & 0.5 \end{pmatrix}$$

The Allyl Model

Verification for the allyl model is more difficult because the total rotation matrix is the result of nine "tuning" operators; it is the product of nine successive rotation matrices. We can check internal consistency of the system by multiplying the eigenvector matrix into the secular matrix (minus x from each principal diagonal element)

$$\begin{pmatrix} 0 & 1 & 0 \\ 1 & 0 & 1 \\ 0 & 1 & 0 \end{pmatrix}$$

to see if the secular matrix is diagonalized thereby and to see if eigenvalues of the allyl system are obtained:

$$\mathbf{AXA} \overset{?}{=} \mathbf{EI} = \mathbf{E}$$

EXERCISE 8-2

Carry out the multiplication above.

Solution 8-2

$$A = \begin{pmatrix} 0.5 & 0.707 & 0.5 \\ 0.707 & 0 & -0.707 \\ 0.5 & -0.707 & 0.5 \end{pmatrix}$$

$$AXA = \begin{pmatrix} 0.5 & 0.707 & 0.5 \\ 0.707 & 0 & -0.707 \\ 0.5 & -0.707 & 0.5 \end{pmatrix} \begin{pmatrix} 0 & 1 & 0 \\ 1 & 0 & 1 \\ 0 & 1 & 0 \end{pmatrix}$$

$$\times \begin{pmatrix} 0.5 & 0.707 & 0.5 \\ 0.707 & 0 & -0.707 \\ 0.5 & -0.707 & 0.5 \end{pmatrix}$$

$$= \begin{pmatrix} 0.707 & 1 & 0.707 \\ 0 & 0 & 0 \\ -0.707 & 1 & -0.707 \end{pmatrix} \begin{pmatrix} 0.5 & 0.707 & 0.5 \\ 0.707 & 0 & -0.707 \\ 0.5 & -0.707 & 0.5 \end{pmatrix}$$

$$= \begin{pmatrix} 2(0.707) & 0 & 0 \\ 0 & 0 & 0 \\ 0 & 0 & 2(-0.707) \end{pmatrix} = \begin{pmatrix} 1.41 & 0 & 0 \\ 0 & 0 & 0 \\ 0 & 0 & -1.41 \end{pmatrix}$$

The matrix **E** that should be found, is found.

Charge Density

Once the eigenvectors have been found, there is much that can be done to transform them into derived quantities that give us an intuitive sense of how HMO calculations relate to the physical properties of molecules that is better than we get from the eigenvalues and eigenfunctions alone. One of these is the charge density. The magnitude of the coefficient of an orbital a_i at a given carbon atom C_i gives the amplitude of the wave function at that atom. The square of the wave function is a probability function; hence, the square of the eigenvector coefficient gives a relative probability of finding the electron. This is the relative charge density too, because a point in the molecule at which there is a high probability of finding electrons is a point of large negative charge density and a portion of the molecule at which electrons are not likely to be found is positively charged relative to the rest of the molecule. Do not forget that, by definition, each *molecular* orbital includes all carbon atoms in the pi electron system. There may be one or two electrons in an orbital ($N = 1$, $N = 2$). Unoccupied orbitals, of course, make no contribution to the charge density.

To obtain the total charge density q_i at atom C_i, we must sum over all occupied or partially occupied orbitals and subtract the result from 1.0, the pi charge density of the carbon atom alone,

$$q_i = 1.0 - \sum Na_i^2 \qquad (8\text{-}12)$$

The sum $\sum Na_i^2$ is the electron probability density at C_i and q_i is the total charge density (positive or negative) relative to the neutral situation.

This definition gives distinctly different charge distributions in, for example, the positively charged ion, free radical, and negatively charged ion of the allyl system. The low-energy orbital for the allyl model has coefficients given by the leftmost column in matrix **A** (Solution 8-1). The positive ion, with two electrons in the lowest orbital has

$$q_1 = q_3 = 1.00 - 2(0.50)^2 = 0.5$$

$$q_2 = 1.00 - 2(0.707)^2 = 0.0$$

Thus the charge in $CH_3CH{=}CH^+$ is not localized at one end of the molecule but (within the Huckel approximations) is concentrated equally at either end,

$$\begin{array}{ccccc} C & - & C & - & C \\ 0.5 & & 0.0 & & 0.5 \end{array}$$

The allyl free radical, with three electrons, has

$$q_1 = q_3 = 1.00 - 2(0.50)^2 - (0.707)^2 = 0.0$$

$$q_2 = 1.00 - 2(0.707) - (0.0)^2 = 0.0$$

Two electrons are in the lowest-energy MO. They give the middle terms in the equations above. The third term arises from the single nonbonding orbital of allyl, middle column, matrix **A**. This leads to

$$\begin{array}{ccccc} C & - & C & - & C \\ 0.0 & & 0.0 & & 0.0 \end{array}$$

as expected for a neutral species.

The negatively charged ion, with four electrons, yields

$$\begin{array}{ccccc} C & - & C & - & C \\ -0.5 & & 0.0 & & -0.5 \end{array}$$

Its charge density distribution is like that of the cation but the sign is reversed.

| EXERCISE 8-3

Write out the charge density diagrams for the positive ion, free radical, and negative ion of the cyclopropenyl system.

Dipole Moments

Knowing the charge density q_i at each atom, one can calculate the dipole moment. First, the charge density at each atom is represented as a vector of length q_i from some arbitrary origin in the direction of the atom in question. If the vector is colinear with a bond and the origin is at an atom, the vector represents a bond dipole moment. All vectors need not represent bond dipole moments. When all charge densities have been represented by vectors, sum the vectors for the total dipole moment.

As an example, take the triply substituted carbon atom in methylene cyclopropene as the origin for the charge densities in that molecule. The charge densities at each atom are -0.478, 0.118, 0.180, and 0.180 according to the numbering in Fig. 8-1. Carbon atom 2 is taken as the origin. The direction of the bond vector from carbon atom 2 to carbon 1 is reversed in the vector diagram because the charge density at carbon 1 is negative. Taking 140 pm (1 pm $= 10^{-12}$ m) as a reasonable value for the C—C bond distance (see below), the vector diagram (Fig. 8-1) shows that the sum of charge vectors is 112 C pm (coulomb picometers). Multiplying by 4.77×10^{-2} to convert to units of debyes, one obtains 5.3 D with the negative end of the dipole at the methylene carbon. This is almost certainly too large. The true value is probably between 1 and 2 D. Nevertheless, an order of magnitude has been calculated and the direction of the dipole is correct. We shall treat more refined dipole moment calculations in the next five chapters.

Bond Orders

Just as it is possible to calculate the electron probability densities at carbon atoms in a pi system, so it is possible to calculate the probability densities *between* atoms. These calculations bear a rough quantitative relationship to the chemical bonds connecting atoms. Because we are calculating only pi electron densities, the results relate only to pi bonds. As the reader may anticipate from the discussion to this point, bonds are not localized between atom pairs in MO theory, but are delocalized over the pi system. One can use either MOBAS or the simple Huckel molecular orbital program SHMO (disk) to obtain the eigenfunctions and eigenvalues for butadiene shown in Fig. 8-2.

Bond orders have been correlated with bond lengths and vibrational force constants. The definition of bond order is

$$P_{ij} = \sum N a_i a_j$$

where N is the number of electrons in the orbital (1 or 2) and a_i and a_j are the coefficients of the bound carbon atoms C_i and C_j. The symbol P_{ij} is given the name *bond order*; it is a measure of the probability of finding

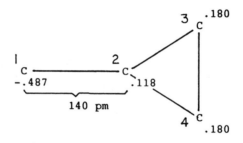

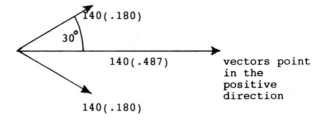

Resultant = 140(.180)cos30 + 140(.180)cos30 + 140(.487)

= 43.6 + 68.2 = 112 C pm

Dipole (see text)

5.3 D

(By convention, the arrow head points to the negative end of the dipole.)

FIGURE 8-1 Dipole moment calculations for methylenecyclopropene.

a pi electron between C_i and C_j. If P_{ij} is large relative to the other bond orders, we anticipate a strong bond. The term *population* will also be used for this symbol later, to denote the expected electron population in an orbital.

For 1,3-butadiene, the order of the 1,2 bond is

$$P_{12} = 2(0.371)(0.600) + 2(0.600)(0.371) = 0.894$$

In the case of the pi bond of ethylene, or any isolated pi bond, the bond order is 1.0. We may take the value 0.894 for the bond order at the 1,2 position as an indication that the 1,2 pi bond in 1,3-butadiene is not exactly the same as the pi bond in ethylene ($P_{12} = 1.00$) but is somewhat

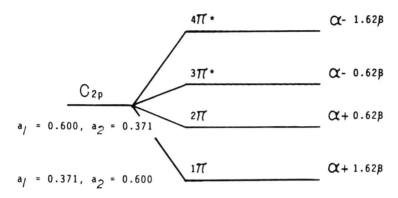

FIGURE 8-2 Energy-level manifold for butadiene.

diminished by delocalization of electrons over the molecular orbital system. Adding the single sigma bond to this result, the "double bond" in 1,3-butadiene is really a 1.89 bond. This delocalization of electrons away from the isolated double bond in the 1,2 position implies an augmentation of the bond order in the 2,3 position; it ought to be more than a single bond by the electron probability density gained from the terminal bonds.

EXERCISE 8-4

Calculate the bond orders for the 2,3 and 3,4 bonds in butadiene. Is the 2,3 bond augmented at the expense of the terminal bonds?

Solution 8-4

$$C \frac{}{1.89} C \frac{}{1.45} C \frac{}{1.89} C$$

Yes.

Delocalization Energy

We have already obtained solutions for the localized system ethylene $E = 2\alpha + 2\beta$, which is the energy of an isolated double bond. In looking at the next more complicated case, allyl, we can regard it as an isolated double bond between two sp^2 carbons to which an sp^3 carbon is attached,

$$C{=}C{-}C$$

or as, in valence bond terminology, a resonance hybrid

$$C{=}C{-}C \leftrightarrow C{-}C{=}C$$

In molecular orbital terminology, the hybrid might be represented by one structure with delocalized pi electrons spread over the sigma-bonded framework

$$\overset{\cdots\cdots\cdots}{C—C—C}$$

These latter two descriptions are equivalent. Chemical evidence leads us to accept either the valence bond or the molecular orbital representation as preferable to the localized representation. One can calculate the eigenvalue (energy) of the delocalized system and the localized system, using program SHMO. The difference between the two is the delocalization energy of allyl. This is the Huckel molecular orbital equivalent of the experimentally determined resonance energy.

EXERCISE 8-5

Write the secular matrix for localized butadiene

$$CH_2{=}CHCH{=}CH_2$$

Solution 8-5

$$\begin{pmatrix} x & 1 & 0 & 0 \\ 1 & x & 0 & 0 \\ 0 & 0 & x & 1 \\ 0 & 0 & 1 & x \end{pmatrix}$$

The matrix is the 1,3-butadiene matrix with the 1s representing atoms C_2 and C_3 omitted to reflect the disconnected nature of the two terminal pi bonds. Another way of looking at the matrix is that it represents two ethylene (localized) pi bonds in the same linear molecule.

EXERCISE 8-6

Calculate the delocalization energies of the positive ion, free radical and negative ion of the allyl model.

Solution 8-6 The energy of the isolated double bond is $2\alpha + 2\beta$. Both ions and the free radical of the allyl system have eigenvalue energies of $2\alpha + 2.828\beta$. The difference is 0.828β in all three cases; hence, the delocalization energies are all 0.828β. The reason the energy is not changed by adding an electron to the allyl positive ion to obtain the free radical is that the electron goes into a nonbonding orbital, which neither augments nor diminishes the energy. The same is true if two electrons are added in to obtain the negative ion. Usage of the term *allyl model* or *allyl system* is illustrated by this exercise. The positive ion, the negative ion, the neutral molecule, and the free radical are all represented by the same energy-level manifold and eigenvectors; they differ only in the number of electrons.

The Free Valency Index

The free valency index F is a measure of reactivity, especially of free radicals:

$$F_r = 1.732 - \sum P_r \qquad (8\text{-}13)$$

where $\sum P_r$ is the sum of bond orders between atom C_i and all other atoms to which it is connected. For example, the free valency index for the terminal carbon atoms in 1,3-butadiene is

$$F = 1.732 - 0.894 = 0.838$$

Within the predictive capabilities of the models, reactivity is given by F_r. The larger F_r, the more reactive the molecule (or ion or radical). Qualitative correlation with experience is seen for some free radicals and alkenes in Fig. 8-3.

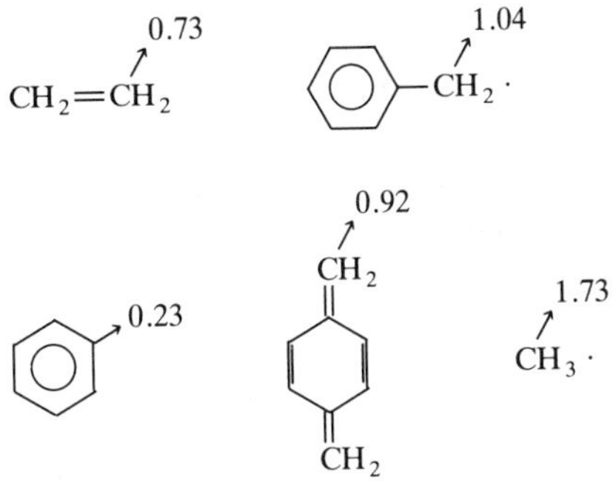

FIGURE 8-3 Free-valency indices.

Resonance Energies

The term *resonance energy* has been used in several ways in the literature, but we shall use it here to mean the difference between the experimentally determined energy of some relatively complicated molecule and the experimental energy expected by analogy to some relatively simple molecule. For example, the enthalpy of hydrogenation of 1-butene is -127 kJ mol^{-1},

$$CH_2{=}CHCH_2CH_3 + H_2 \longrightarrow CH_3CH_2CH_2CH_3$$

from which we can extrapolate the value of $2(-127) = -254$ kJ mol^{-1} for hydrogenation of 1,3-butadiene,

$$CH_2{=}CHCH{=}CH_2 + 2H_2 \longrightarrow CH_3CH_2CH_2CH_3$$

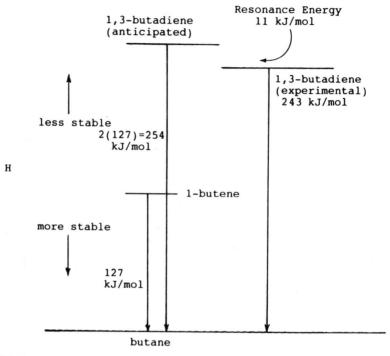

FIGURE 8-4 Enthalpy-level diagram for 1-butene and 1,3-butadiene. Arrows point down because ΔH_h for hydrogenation is negative (exothermic).

The actual value of the enthalpy of hydrogenation of 1,3-butadiene is -243 kJ mol^{-1}. Both are hydrogenated to the same product; hence, the enthalpy diagram (Fig. 8-4) shows that 1,3-butadiene is 11 kJ mol^{-1} lower in enthalpy than it "ought" to be on the basis of the reference standard, 1-butene. Changing the reference standard to ethylene, for example, gives a different value for the resonance energy. Ethylene has an enthalpy of hydrogenation of -136.0 kJ mol^{-1}, which leads to $2(-136) = -272$ kJ mol^{-1} as the anticipated enthalpy of hydrogenation of 1,3-butadiene. The experimental value is 29 kJ mol^{-1} less exothermic.

In a somewhat loose terminology, 1,3-butadiene is said to be *stabilized* relative to the reference standards. Valence-bond theory explains this stabilization using resonance structures (Wheland, 1955); hence, it is called resonance stabilization. The difference between the enthalpy of hydrogenation anticipated on the basis of a single reference structure and the experimental enthalpy of hydrogenation is the *resonance energy*. There are other ways to determine resonance energies (Wheland, 1955; Streitwieser, 1961).

EXERCISE 8-7

Turner (see Jensen, 1976) has measured ΔH_h of cyclohexene and found it to be -113 kJ mol^{-1}. Using cyclohexene as the reference standard, calculate the resonance energies of 1,3-cyclohexadiene ($\Delta H_h = -224$ kJ mol^{-1}), 1,4-cyclohexadiene ($\Delta H_h = -225$ kJ mol^{-1}), and benzene ($\Delta H_h = -216$ kJ mol^{-1}). Comment upon these results. The value for benzene (Conant and Kistiakowsky, 1937) has been corrected to the experimental conditions of Turner's results.

Extended Huckel Theory: Wheland's Method

One restriction imposed by Huckel theory that is rather easy to release is that of zero overlap for nearest-neighbor interactions. One can retain $\alpha - E$ as the diagonal elements in the secular matrix and replace β by $\beta - ES$ as nearest-neighbor elements where S is the overlap integral. Now,

$$\begin{pmatrix} \alpha - E & \beta - ES & 0 & \cdots \\ \beta - ES & \alpha - E & \beta - ES & \cdots \\ \cdots & & & \\ \cdots & 0 & \beta - ES & \alpha - E \end{pmatrix} \tag{8-14}$$

is the secular matrix.

We can make the substitution

$$x = \frac{\alpha - E}{\beta - ES} \tag{8-15}$$

which causes the secular matrix to take the same form that it did in the simple Huckel theory,

$$\begin{pmatrix} x & 1 \\ 1 & x \end{pmatrix}$$

for ethylene. From the definition of x,

$$E = \frac{\alpha - x\beta}{1 - xS} \tag{8-16}$$

Mullikan has shown that the overlap integral for hydrogen-like p orbitals in linear hydrocarbons is about 0.27 (Mullikan, 1949). Hydrogen-like orbitals used as basis functions for molecular orbital calculations are called Slater-type orbitals (STOs).

EXERCISE 8-8

Prove Eq. (8-16) from the definition of x.

EXERCISE 8-9

What is the energy separation $E^* - E$ of the bonding and antibonding orbitals in ethylene, assuming that the overlap integral S is 0.27.

Solution 8-9

$$E = \frac{\alpha - \beta}{0.73}$$

$$E^* = \frac{\alpha + \beta}{1.27}$$

$$\beta = -1, \qquad \alpha = 0$$

$$E = -0.79, 1.37$$

The separation is 2.13β.

Extended Huckel Theory: Hoffmann's Method

Hoffmann's extension (Hoffmann, 1963) of the Huckel theory includes all bonding orbitals in the secular matrix rather than just all pi bonding orbitals. This inclusion increases the complexity of the calculations to the point that they are not practical without a computer.

The basis set is the LCAO set of valence orbitals

$$\psi_i = \sum a_{ij}\phi_i$$

but even for ethylene, this leads to a 12 × 12 secular matrix because there are four valence electrons on each of two carbon atoms and one on each of the four hydrogen atoms.

The orbitals used are the $1s$ AOs of hydrogen and the $2s$ and $2p$ AOs of carbon. These are assumed to be of the same form as the hydrogen-like orbitals, that is, STOs. Considerable limitation on the number of linear combinations of STOs is possible because of symmetry properties of the spin states of the MOs produced. Such orbitals are said to be antisymmetrized, because spatially symmetric orbitals can be combined only with antisymmetric spin functions and vice versa (Hinchliffe, 1988).

We shall now fill the secular matrix **H** with elements H_{ij} over the entire set of valence orbitals. The diagonal elements are

$$H_{ii} = \begin{cases} -13.6 & 1s & \text{hydrogen} \\ -21.4 & 2s & \text{carbon} \\ -11.4 & 2p & \text{carbon} \end{cases}$$

which are the ionization energies in electron volts. The energies are negative as always for bound states relative to an arbitrary zero of energy defined as the energy of the unbound state. (See Computer Project 5-3 for determination of the ionization energy of hydrogen. Convert the answer to electron volts.)

Off-diagonals are given by

$$H_{ij} = 0.88(H_{ii} + H_{jj})S_{ij}$$

where S is, once again, the overlap integral. This is a kind of average of H_{ii} and H_{jj} modified by the overlap and another empirical factor 0.88.

The S_{ij} are geometry dependent. One can vary the geometry of the molecule and select the structure that gives the lowest energy, thereby obtaining the best geometry from the alternatives tried. This is a concept that will be developed to a sophisticated level in the remaining chapters of this book.

Applications of extended Huckel theory (EHT) include calculation of the rotational barrier in ethane and of the chair–boat conformational energies in cyclohexane. It has been largely supplanted by *ab initio* and semiempirical calculations; therefore, it will not be used in the computer projects.

The Programs

MOBAS was written by the author (Rogers, Angelis, and Mc Lafferty, 1983) in BASIC to illustrate matrix inversion in molecular orbital calculations. It is modeled after a program in FORTRAN II given by Dickson (1968).

SHMO is a simple Huckel MO program in FORTRAN that functions much as MOBAS does and is also based on the Dickson program. SHMO is in compiled FORTRAN and source code on the accompanying disk. Compiled SHMO is in executable code (called SHMO.EXE on the disk). For IBM compatible computers, all that need be done is to type in SHMO from the system level (do not go into BASIC)

> SHMO

The program responds with a series of prompts. The input format is similar to MOBAS. Matrix elements are entered row, column, element, 0, except for the last, which ends in 99. After receiving the answers, the program prints out the eigenvalues and eigenfunctions for the problems presented to it. Note that data are input to the program from the keyboard and not built into it, as in MOBAS.

If you wish to modify SHMO or run it on a machine that is in any way different from the machine that it was compiled on, you will have to recompile. Compiling FORTRAN programs is outside the scope of this book but is described in detail in the manuals available with any commercial compiler. SHMO was compiled using IBM FORTRAN Compiler 2.0.

HMO is a more elaborate Huckel MO program that calculates charge densities, bond orders, and free valency indices. It is written in FORTRAN and is a modification of a program by Greenwood (1972). HMO must also be recompiled for different machines. A source and executable version are available on the enclosed disk.

ICON8 is an extended Huckel program of the Hoffmann type distributed by Quantum Chemistry Program Exchange (QCPE 344). It carries out EHT calculations on up to 50 atoms containing *s*, *s*, and *p*, or *s*, *p*, and *d* electrons. We will not use it in the exercises.

COMPUTER PROJECT 8-1 | Calculations Using SHMO

SHMO, being in compiled FORTRAN, is much faster than MOBAS, which is written in standard interpreted BASIC. Solve the Huckel matrix for the hexatriene model using MOBAS and again using SHMO. Record the run time for this problem with each program. Record the times obtained using your computer.

Because of its advantage in speed, we can study larger models using SHMO than we can with MOBAS. Use SHMO to obtain the energy-level diagrams for the models methylenepentadiene, bicyclohexatriene, and styrene:

Polarographic oxidation entails removing one or more electrons from the species undergoing oxidation at a mercury or similar electrode. The more tightly held an electron is, the more difficult it is to remove; hence, the higher the electrode potential necessary to remove it. Make the reasonable hypothesis that the electron removed in a one-electron oxidation comes from the highest occupied orbital (HOMO). Using SHMO, determine the HOMO for benzene, biphenyl, and naphthalene. Note that all of the occupied orbitals in these molecules have negative energies, i.e., they are below α by an energy measured in units of β. The least negative HOMO has the highest energy electrons and is most easily oxidized. Arrange the three compounds in order of increasing oxidation potential:

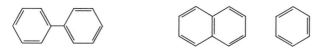

COMPUTER PROJECT 8-2 | *Dipole Moments*

Using Program HMO, calculate the dipole moment of methylenecyclopropene by the HMO method. The program gives total charge densities at each carbon atom, making the calculation of dipole moments essentially a geometric problem. The single charge of the carbon atom must be subtracted from the total charge density to obtain the charge density used in the dipole moment calculation, that is, a computer output at carbon 1 of 1.5 leads to a pi charge density at that atom of 0.5. Calculate the dipole moment of fulvene by the HMO method. Assume, for the calculation, that the endocyclic double bonds are parallel as in the diagram and that the angle at carbon atom 2 is the same as in methylenecyclopropene:

Methylenecyclopropene   Fulvene

These assumptions are not true but we will be able to arrive at accurate geometries in Chapters 10 through 13.

In which direction is the dipole moment of methylenecyclopropene? In which direction is the dipole moment of fulvene?

Program HMO gives you the option of modifying one or more of the elements input to the semimatrix. Calculate the charge densities of cyclopropenone by entering the semimatrix

$$
\begin{pmatrix}
2 & & & \\
1 & 0 & & \\
0 & 1 & 0 & \\
0 & 1 & 1 & 0
\end{pmatrix}
$$

and selecting the modification option by typing 0001 at the prompt: *Enter number of elements to be modified I4 format.* Respond to the prompt: *Enter row (I2), column (I2), and new element (F6.3)* by typing

02011.414

This leads to the semimatrix

$$\begin{pmatrix} 2 & & & \\ 1.414 & 0 & & \\ 0 & 1 & 0 & \\ 0 & 1 & 1 & 0 \end{pmatrix}$$

for cyclopropenone. Program HMO automatically loads the full matrix from the semimatrix because Huckel molecular orbital matrices are always symmetrical; hence, the program "knows" what the elements are above the principal diagonal. Calculate the dipole moment of cyclopropenone.

Use the same method to calculate the dipole moment of cyclopentadienone. Assume, for the calculation, that the endocyclic double bonds are parallel and the angle at carbon 2 is the same as in cyclopropenone:

Cyclopropenone   Cyclopentadienone

There is a substantial difference in dipole moments between methylenecyclopropene and cyclopropenone but the difference between fulvene and cyclopentadienone is much smaller. Explain.

COMPUTER PROJECT 8-3 | *Conservation of Orbital Symmetry*

Conservation of orbital symmetry is a general principle that requires orbitals of the same phase (sign) to match up in a chemical reaction. For example, if terminal orbitals are to combine with one another in a cyclization reaction as in pattern A, they must rotate in the same direction (conrotatory overlap), but if they combine according to pattern B, they must rotate in opposite directions (disrotatory). In each case, rotation takes place so that lobes of the pi orbitals, which are of the same sign, overlap.

Pattern A:

Pattern B:

Obtain the orbitals of butadiene and predict whether the cyclization of butadiene to cyclobutene is conrotatory or disrotatory:

Perform the same calculation for 1,3,5-hexatriene:

Conrotatory and disrotatory concerted reactions can often be distinguished by chemical means. For example, using the results of the previous calculation, predict whether the cyclizations of 2,4-hexadiene will lead to the *cis* or *trans* dimethyl cyclobutene

Perform the same calculation for cyclization of 2,4,6-octatriene. Which isomer of dimethylcyclohexadiene is formed?

PROBLEMS | *Chapter* 8

1 Find the determinant of

$$\begin{pmatrix} \cos\theta & \sin\theta \\ \sin\theta & -\cos\theta \end{pmatrix}$$

2 Determine the eigenvectors and eigenvalues for methylenecyclobutene.

3 Determine the delocalization energy for methylenecyclobutene.

4 Determine the dipole moment for methylenecyclobutene.

5 Refer to Computer Project 7-2. Calculate β in units of electron volts using Wheland's extension of Huckel molecular orbital theory.

6 Print program MOBAS. Identify the statements in the program that generate the eigenvector matrix **A** by performing the same rotation operations on **I** that it performs on the input matrix in order to generate the eigenvalue matrix.

7 Determine the dipole moment of cyclobutenone.

8 Spectroscopically determined values of β vary, but are usually around -2.4 eV (Exercise 7-2). In the section on resonance stabilization, we saw that thermodynamic measurements of the total resonance stabilization of butadiene yield 11 and 29 kJ mol^{-1} according to the standard chosen. Calculate the delocalization energy of 1,3-butadiene in units of β. Determine two values for the "size" of the energy unit β from the two thermochemical estimates given. Do these agree well or poorly with the spectroscopic values?

9 The delocalization energy of benzene is 2β (verify this). From information in Exercise 8-7, calculate yet another value for the "size" of the β unit based on the thermodynamic values of the enthalpy of formation of benzene. Does this value agree with those in Problem 8-8? Does it agree with the spectroscopic values?

REFERENCES

Conant, J. B. and Kistiakowsky, G. B., 1937. *Chem. Rev.*, **20**, 181.

Coulson, C. A., O'Leary, B., and Mallion, B. B., 1978. *Huckel Theory for Organic Chemists*. Academic, New York.

Dickson, T. R., 1968. *The Computer and Chemistry*. Freeman, San Francisco, CA.

Greenwood, H. H., 1972. *Computing Methods in Quantum Organic Chemistry*. Wiley Interscience, New York.

Hinchliffe, A., 1988. *Computational Quantum Chemistry*. Wiley, New York.

Hoffmann, R., 1963. *J. Chem. Phys.* **39**, 1397.

Jensen, J. L., 1976. *Prog. Phys. Organic Chem.*, **12**, 189.

Mulliken, R. S., et al., 1949. *J. Chem. Phys.*, **17**, 1248.

Rogers, D. W., Angelis, B. P., and Mc Lafferty, F. J., 1983. *Am. Lab.*, **15** (11), 54.

Streitwieser, A., 1961. *Molecular Orbital Theory for Organic Chemists*. Wiley, New York.

Wheland, G. W., 1941. *J. Amer. Chem. Soc.*, **63**, 2025.

Wheland, G. W., 1955. *Resonance in Organic Chemistry*. Wiley, New York.

9 Self-Consistent Field Theory

Because of its severe approximations, the Huckel method tends to gloss over most of the real problems of molecular orbital theory. This is not to say that Huckel did not see the problems clearly. Indeed, it was his great achievement to circumvent each stumbling block with a stroke so bold as to seem reckless, yet to retain the essentials of the theory intact. Huckel theory provided the foundation and stimulus for a generation's research, most notably in organic chemistry.

To advance beyond Huckel's method, we must first look at the problem of calculating the elements of the secular matrix. This is a problem Huckel simply swept away by calling the diagonal elements α and all others β. He further simplified the problem by setting $\alpha = 0$ and giving β the simplest possible definition: $\beta = 1$ for adjacent carbon atoms and $\beta = 0$ for nonadjacent carbons (Chapter 7). Huckel molecular orbital energies come out in units of β relative to the arbitrary zero point α. Attempts to evaluate β in units of electron volts or kilojoules per mole do not, however, give consistent values (Problem 8-8).

ELEMENTS OF THE SECULAR MATRIX

To begin a more general approach to molecular orbital theory, we shall describe a variational solution of the prototypical problem of the ground-state energy of the electron in a box (Mc Quarrie, 1983). The particle in a one-dimensional box has an exact solution

$$E = \frac{n^2 h^2}{8ma^2} \tag{9-1}$$

Let the dimension of the box be 1 in units consistent with the units of Planck's constant h and consider only the ground state, for which $n = 1$,

$$E = \frac{h^2}{8m} = 0.125 \left(\frac{h^2}{m} \right) \tag{9-2}$$

The wave function is also known,

$$\psi = A \sin(kx) \tag{9-3}$$

but we wish to approximate it by a linear combination of functions,

$$\psi = c_1 x(1 - x) + c_2 x^2 (1 - x)^2 \tag{9-4}$$

Call this combination

$$\psi = c_1 f_1 + c_2 f_2 \tag{9-5}$$

There are two functions, so we shall obtain two molecular eigenvalues; the ground-state energy will be the lower of the two. The full secular matrix is

$$\begin{pmatrix} H_{11} - ES_{11} & H_{12} - ES_{12} \\ H_{21} - ES_{21} & H_{22} - ES_{22} \end{pmatrix} = 0 \tag{9-6}$$

If we do not make any simplifying assumptions, we must calculate the matrix elements

$$H_{11} = \int f_1 H f_1 \, d\tau$$

$$H_{12} = \int f_1 H f_2 \, d\tau$$

$$H_{21} = \int f_2 H f_1 \, d\tau$$

$$H_{22} = \int f_2 H f_2 \, d\tau$$

$$\tag{9-7}$$

where the hamiltonian operator for a particle in a box is

$$\frac{-\hbar^2}{2m} \frac{d^2}{dx^2} \tag{9-8}$$

In this problem, the integral over "all space" $d\tau$ is in only one dimension, x. The limits of integration are the dimensions of the box, 0 and 1, in whatever unit was chosen.

EXERCISE 9-1

Calculate matrix element H_{11}.

Solution 9-1

$$f_1 = x(1 - x)$$

$$\frac{d^2}{dx^2}(x(1 - x)) = -2$$

$$\frac{-\hbar^2}{2m} \int x(1 - x) \frac{d^2}{dx^2} x(1 - x) \, d\tau$$

$$= \frac{-\hbar^2}{2m} \int x(1 - x)(-2) \, d\tau \tag{9-9}$$

$$= \frac{\hbar^2}{m} \left(\frac{x^2}{2} - \frac{x^3}{3} \right) \Big|_0^1$$

$$= \frac{\hbar^2}{m} \left(\frac{1}{6} \right) = \frac{\hbar^2}{6m}$$

Also, one needs to calculate the matrix elements

$$S_{11} = \int f_1 f_1 \, d\tau$$

$$S_{12} = \int f_1 f_2 \, d\tau$$

$$S_{21} = \int f_2 f_1 \, d\tau$$ (9-10)

$$S_{22} = \int f_2 f_2 \, d\tau$$

Again, the limits of integration are 0 to 1.

EXERCISE 9-2

Calculate S_{11}.

Solution 9-2

$$S_{11} = \int x(1-x)x(1-x) \, d\tau$$

$$= \int (x^2 - x^3 - x^3 + x^4) \, d\tau$$

$$= \tfrac{1}{3} - \tfrac{1}{4} - \tfrac{1}{4} + \tfrac{1}{5}$$

$$= \tfrac{1}{30}$$

Now, calculating all H and S elements in the same way, and inserting them into the secular matrix in the proper places,

$$\begin{vmatrix} \dfrac{\hbar^2}{6m} - \dfrac{E}{30} & \dfrac{\hbar^2}{30m} - \dfrac{E}{140} \\[2mm] \dfrac{\hbar^2}{30m} - \dfrac{E}{140} & \dfrac{\hbar^2}{105m} - \dfrac{E}{630} \end{vmatrix} = 0 \qquad (9\text{-}11)$$

Dividing each element by $\hbar^2/m$ and setting

$$x = \dfrac{E}{\hbar^2/m}$$

yields

$$\begin{vmatrix} \dfrac{1}{6} - \dfrac{x}{30} & \dfrac{1}{30} - \dfrac{x}{140} \\[2mm] \dfrac{1}{30} - \dfrac{x}{140} & \dfrac{1}{105} - \dfrac{x}{630} \end{vmatrix} = 0 \qquad (9\text{-}12)$$

This can be cleared of fractions by multiplying by 1260 to obtain

$$\begin{pmatrix} 210 - 42x & 42 - 9x \\ 42 - 9x & 12 - 2x \end{pmatrix} = 0 \qquad (9\text{-}13)$$

These are the equations [Eqs. (1-10)] that were solved iteratively in Computer Project 1-3 to yield the roots

$$4.93487 \quad \text{and} \quad 51.065$$

The smaller of the two roots is the one we seek for the ground-state energy of the system.

Thus,

$$x = \frac{E}{\hbar^2/m}, \qquad E = \frac{x\hbar^2}{m}$$

and, recalling that $\hbar = h/2\pi$,

$$E = \frac{4.93487}{(2\pi)^2} \left(\frac{h^2}{m} \right) \qquad (9\text{-}14)$$

$$= 0.125002 \left(\frac{h^2}{m} \right) \qquad (9\text{-}15)$$

as contrasted to the exact solution of $0.125(h^2/m)$. Note that the energy obtained from the approximate solution is slightly larger than the solution obtained from the exact wave function.

One of the things illustrated by this calculation is that a surprisingly good approximation to the eigenvalue can often be obtained from a combination of functions that do not represent the exact eigenfunction very closely. That is, the eigenvalues are not very sensitive to the eigenfunctions. This is one reason why the LCAO approximation and Huckel theory in particular work as well as they do.

Another feature of advanced molecular orbital calculations that we can anticipate from this simple example is that calculating matrix elements for real molecules can be a formidable task.

SEMIEMPIRICAL VS. AB INITIO

If we are willing to substitute numerical values, obtained from empirical observations, into the secular matrix we can avoid calculating matrix elements entirely. Filling in the matrix elements by fitting them to spectroscopic or other experimental data leads to a semiempirical calculation. Such methods are not fully empirical because they are rooted in the theory of the Schroedinger equation as expressed through the variational principle.

If no empirical infusion of information takes place and physical results (energies, charge distributions, etc.) are derived entirely by calculating matrix elements from linear combinations of orbitals (atomic or other), the results are said to be derived from the beginning or *ab initio*.

The remainder of this chapter is concerned with a semiempirical method, the self-consistent field (SCF) method, that was developed by Pariser and Parr (1953) and by Pople (1953) and usually goes under the name of the PPP method. Numerical values are obtained from experimental observations, substituted for the hamiltonian integrals H_{ij} and solved as in the Huckel method. In addition, empirical electron repulsion parameters will be put into the matrix and an iterative scheme will be set up so that the solution converges (may converge) on self-consistency.

PPP SELF-CONSISTENT FIELD CALCULATIONS

In the Huckel method, we assumed an initial constant pi electron density q_1 about all carbon atoms in a conjugated pi electron system of one electron per carbon. We also took the electron exchange integrals between atoms to be 1 arbitrary unit of energy, according to whether the atoms are connected ($\beta = 1$) or not connected ($\beta = 0$). The eigenvectors (coefficients) generated in diagonalizing the secular determinant, however, yield electron densities and bond orders that are in contradiction to the original assumptions. In particular, if a bond order between atoms r and s is large and that between u and v is small, β is not the same for these atom pairs, but $\beta_{rs} > \beta_{uv}$.

We also took the overlap integrals S_{12} to be 0 in Chapters 7 and 8. In the Huckel theory, this has little effect on the calculated energy. In more advanced theories, however, the manner of handling overlap will be significant. We shall start with the approximation of zero differential overlap (ZDO) within the constraint of the pi-sigma separation (Chapter 7). We shall look into ways of releasing these constraints in later chapters.

It seems reasonable that, by taking into account the new information we have generated from the first set of calculated eigenvectors, we can correct the information going into the calculations and obtain a better result on the second try. If this works once, it should work many times. There may be convergence to a result which, though not exact, is self-consistent and is a better description of the molecule than the single matrix diagonalization of the Huckel method. This is the essence of the *self-consistent field* (SCF) method.

We have the makings of an iterative computer method: Start by assuming values for β, calculate bond orders, modify β according to the results of the bond order calculations, rediagonalize using the new β values to get new bond orders, and so on, until the results of one iteration are not measurably different from those of the last. The results are then self-consistent.

The Schroedinger equation has, of course, the same form,

$$H^{SCF}\psi_i = E_i^{SCF}\psi_i \tag{9-16}$$

where H and E are, as in Huckel theory, approximate. Also, we can let

$$\psi_i = \sum a_j\phi_j \tag{9-17}$$

where ϕ_j are, in the LCAO approximation, the atomic orbitals of all electrons involved in the pi bond structure of the molecule. This leads to a set of secular equations

$$\sum \left(H_{rs}^{SCF} - E_i^{SCF}S_{rs}\right)a_{is} = 0 \tag{9-18}$$

and a secular matrix similar in form to the Huckel matrix, for example,

$$\begin{pmatrix} \alpha^{SCF} & \beta^{SCF} \\ \beta^{SCF} & \alpha^{SCF} \end{pmatrix} = 0 \tag{9-19}$$

for ethylene. (Henceforth vector and matrix boldface notation will be omitted except where essential to the argument.)

First, we shall consider only the simplest class of SCF calculations, that of conjugated pi systems in alternant hydrocarbons like ethylene and butadiene. In contrast to simple Huckel theory, the SCF coulomb integral is defined by

$$\alpha^{SCF} = \omega_r + \tfrac{1}{2}P_{rr}y_{rr} + \sum \left[P_{ss}y_{rs} - \left\langle r|V_s|r\right\rangle\right] \tag{9-20}$$

where ω_r is the ionization energy of an electron at carbon atom r, influenced by the sigma framework potential energy V_s, P_{rr} and P_{ss} are electron densities on carbon atoms r and s, and y is an electron repulsion integral. (For a more detailed discussion of repulsion integrals, see Chapter 12.) The term $(r|V_s|r)$ is the potential energy of an electron centered on atom r in the field of the ion consisting of all nuclei and electrons except that electron. For alternant hydrocarbons, one can set $\alpha = 0$ for the first step.

An equation for modifying α from one iteration to the next is written, which is a simplified form of Eq. (9-20):

$$\alpha_{new}^{SCF} = \alpha_r^{SCF} + \tfrac{1}{2}P_{rr}y_{rr} + \sum (P_{ss} - 1)y_{rs} \tag{9-21}$$

for carbon atom r. See Greenwood (1972) for the rationale of substituting the fitted parameter y_{rs} for $\langle r|V_s|r\rangle$. The parameter β_{rs} may be estimated for the first iteration from spectroscopy.

An equation is written for modifying β from one iteration to the next,

$$\beta_{new}^{SCF} = \beta_{rs}^{SCF} - \tfrac{1}{2}P_{rs}y_{rs}, \qquad r = s \tag{9-22}$$

for carbon atoms r and s, where y again designates electron repulsion integrals over all space and P_{rs} is the electron density between atoms. This is the bond order for adjacent carbon atoms. Note, however, that β values are also calculated for nonadjacent carbon atoms in this procedure. In this more general context, β is called the *framework resonance integral*.

In Eqs. (9-20) and (9-21), P_{rr} is the electron density on atom r so that $P_{rr}y_r$ is an electron repulsion energy, which is large if the electron density at r is large and small if the electron density is small. The third term on the right of Eq. (9-20) is the sum over all $s = r$ of potential energy on an electron at r due to the field caused by a nucleus at position s and repulsion due to the electron density between r and s, y_{rs}.

The parameter y is assigned an empirical value. By one scheme, y_{rr} is taken to be the ionization energy of carbon atom r (Computer Project 5-3). More generally, a physical model of interacting negatively charged spheres is used as the basis of the calculation of repulsive energies and the results are fitted to conform with experimental measurements.

Pariser and Parr adjusted the necessary parameters to the empirical singlet and triplet excitation energies in benzene to obtain

$$y_{11} = 11.35 \text{ eV}$$

$$y_{12} = 7.19 \text{ eV}$$

$$y_{13} = 5.77 \text{ eV}$$

$$y_{14} = 4.97 \text{ eV}$$

where the subscript 12 indicates nearest neighbors whereas 13 and 14 are the next most distant carbon atoms.

Fitting β to the HOMO–LUMO transitions in benzene in a manner similar to Computer Project 7-2 yields

$$\beta = -2.37 \text{ eV}$$

For alternant hydrocarbons, $P_{ss} = 1$; hence, the last term in Eq. (9-21) drops out. We shall simplify notation by letting a primed value indicate the "new" value of the SCF matrix. This leads to

$$\alpha' = \alpha + \tfrac{1}{2}P_{rr}y_{rr}$$
$$\beta' = \beta - \tfrac{1}{2}P_{rs}y_{rs}$$

(9-23)

One final point: Because β values are symmetrically distributed about a single energy, one can complete the analogy between Huckel theory and PPP theory by selecting that energy as α and designating $\alpha = 0$ at each iterative step, not just the first. Now energies take the form $E = \alpha \pm \beta$ as in Huckel theory, although the calculated values of β are not the same. Energy distribution is not symmetrical for molecules other than alternant hydrocarbons.

Ethylene

The simplest application is to ethylene. The SCF matrix is

$$\begin{pmatrix} \alpha^{SCF} - E & \beta^{SCF} \\ \beta^{SCF} & \alpha^{SCF} - E \end{pmatrix}$$

We can calculate β^{SCF} for the first and only diagonalization

$$\beta^{SCF} = \beta_{rs} - \tfrac{1}{2}P_{12}y_{12}$$

where β_{rs} is the empirical value, in electron volts, just mentioned and y_{12} is the repulsion integral for neighboring atoms, i.e., 7.19 for neighboring carbon atoms, which are the only kind in ethylene. The bond order in ethylene is 1.00 and does not change with the calculation; hence,

$$\beta^{SCF} = -2.37 - \tfrac{1}{2}(1.00(7.19))$$

$$= -5.96 \text{ eV}$$

The form of the SCF matrix is the same as the Huckel matrix; hence, we may substitute $x = (\alpha^{SCF} - E)/\beta^{SCF}$ to obtain

$$\begin{pmatrix} x & 1 \\ 1 & x \end{pmatrix} = 0$$

which is diagonalized and leads to

$$E = \alpha^{SCF} \pm \beta^{SCF}$$

as in the Huckel theory except that $\beta^{SCF} = -5.96$ eV.

The solution comes out to be very similar to the Huckel solution for ethylene: There are two energy levels equidistant from the defined zero energy, one energy level 5.96 eV above, and one energy level 5.96 eV below the reference level:

The pi electron bonding energy for the doubly occupied pi orbital in ethylene is calculated as 9.86 eV.

In the case of conjugated systems larger than ethylene, substitution of α' and β' according to Eqs. (9-23) yields a new and lower energy and eigenvalues that are different from the previous diagonalization. New eigenvalues mean that a new charge density P_{rr} and a new P_{rs} can be calculated. These are multiplied by the empirical constants y_{rr} and y_{rs} in Eqs. (9-23) to obtain α'' and β''. This iterative method is continued until the results agree with one another to within some specified convergence criterion. After convergence, solutions are *self-consistent*.

DETERMINANTAL WAVE FUNCTIONS

The complete wave function must contain provision for antisymmetric exchange of all electrons. Because single-electron wave functions are approximate solutions to the Schroedinger equation, one would expect any linear combination thereof to be an approximate solution also. For more than a few basis functions, the number of possible linear combinations can be very large. Fortunately, spin and the Pauli exclusion principle reduce this complexity. Thus far, we have considered only the space part of one-electron orbitals, but each orbital also has a spin part. Moreover, the Pauli principle says that no two one-electron orbitals may be identical in all respects. In particular, if the space parts are identical, the spin parts must be opposite in sign $(+\frac{1}{2}, -\frac{1}{2})$. Determinants have the property that when rows or columns are exchanged (Exercise 4-6), the sign of the determinant changes sign. Slater showed that basis functions arrayed as a determinant change sign in the spin part so as to obey the Pauli principle. Linear combinations that do not arise on expansion of such a determinant, called a Slater determinant, do not obey the Pauli principle and can be discarded. Wave functions that do satisfy the Pauli principle are acceptable linear combinations arising from the Slater determinant and are said to be antisymmetrized. It is possible to make a linear combination of antisymmetrized determinantal wave functions that are approximate solutions of the Schroedinger equation. We shall call these *multiple-determinant wave functions*. By the general principle that larger basis sets more closely approximate the Schroedinger equation than smaller subsets thereof, multiple-determinant wave functions should be better approximations to the solution of the Schroedinger equation than single-determinant wave functions. In practice, multiple-determinant wave functions in PPP theory do not have much effect on the ground state of the molecule, but they do affect the excited states; hence, multiple-determinant wave functions are important in determining spectra that involve the transition of an electron from the ground state to an excited state. At present, the use of multiple-determinant wave functions is limited by computer size and speed.

Configuration Interation

As described, the SCF procedure overestimates electronic repulsion because electrons are treated as though they move in an average potential and no account is taken for the tendency of two electrons in the same orbital to avoid one another because of the mutual repulsion of their negative charges (charge correlation). Molecular orbital calculations using a single determinant, like those discussed to this point, give a value for the energy that is too large by an amount called the *configuration energy*. This error can be diminished by use of multiple-determinant wave functions.

These wave functions are linear combinations of single-determinant functions. Each single-determinant wave function has some unoccupied orbitals called *virtual* orbitals. Replacing a filled orbital with a virtual orbital gives a new basis function for the linear combination that is to generate the multiple-determinant wave function. The multiple-determinant wave function has a smaller correlation energy than the single-determinant function; the error due to electron correlation is smaller. More than one substitution of virtual for occupied orbitals can be made, approaching the full configuration solution, called a CI solution. Limited CI calculations are common. The degree of CI substitution chosen is a trade-off between accuracy required and computer time allowed because calculations of full CI interactions are time-consuming.

The Programs

Two SCF programs were used. One is a modified and cut-down version of the suite of SCOF programs given by Greenwood (1972). It is provided on the accompanying diskette. The diskette contains two versions: SCF.EXE, for use with machines that are compatible with the Tandy 1000 SX, and SCF.FOR, the source program, based on the programs given by Greenwood. Those who wish to use SCF on a different machine or to make modifications, must recompile. SCF.EXE was compiled using IBM FORTRAN compiler 2.0. The modified version of SCF provided here does not carry out CI calculations.

A full SCF–CI program for mainframe is available as program QCPE 314. There is a fairly elaborate input protocol (Greenwood, 1972), which is described in the documentation accompanying the QCPE program.

The second PPP–SCF–CI program is a simplified version by Griffiths, Lasch, and Schermaier (program QCMPE 054) that is available from QCPE on diskette. Note that programs designated QCPE are for mainframe computers and those designated QCMPE are for microcomputers. The disk contains the FORTRAN source file, an executable file for IBM-PC and compatible microcomputers, a sample input file, and a sample output file for comparison with the user's run on the demonstration input file. The program uses an input file consisting of geometric information and ionization potentials, which it converts to repulsion integrals and which it uses in establishing an initial matrix for diagonalization and modification by the procedure given with Eqs. (9-23). The CI portion of the program calculates and prints out the absorption spectrum in terms of a series of absorptions with spectral intensities and wavelengths, in nanometers.

The input protocol is quite different from what we have seen so far and will be described as part of Computer Program 9-3. Configuration interaction causes discrepancies between the output and that of the earlier PPP treatment. Even at the SCF level, without CI, the results are not the same because the electron repulsion integrals are different. The Griffiths program was specifically designed to calculate absorption spectra in the visible region, which it does very well. In designing a semiempirical program to perform some specific function, one optimizes the empirical parameters to reproduce certain selected experimental results from that field, in this case, spectra of dyestuffs. A program parametrized to do one thing should not be expected to do another, unrelated, thing equally well.

EDLIN Edit

Some editor is a necessity for creating and modifying input files. Editor EDLIN comes as a supplementary program to the MS-DOS 3.2. Whatever your input file is, we shall call it INPUT. Copy your demonstration input file from Program PPP to a backup with the command COPY INPUT INPUT.BAK or the equivalent. To enter editor EDLIN, put the *supplementary programs* disk into drive A as the master drive and the PPP disk, for example, into drive B. At A > type EDLIN B:INPUT. At the edit prompt *, you are ready to edit file INPUT according to the system documentation. Deletions in the input file are achieved using n*d* where **n** is the line number to be deleted. Lines are called up for alteration by entering the line number **n**; changes are made by moving the cursor and striking over; a line or lines before **n** are inserted using the edit command n*i*. To retain a symbol in a line, move the cursor over it but do not strike over. Do not worry if you ruin your first editing job; you can always recopy the backup. Exit the editor by typing **e** at the edit prompt *.

COMPUTER PROJECT 9-1 | SCF Calculations of Ultraviolet Spectral Peaks

This project will familiarize you with the input necessary to carry out calculations using program SCF. The concept involved is the reverse of that used in Computer Project 7-2. In the earlier project, you were asked to fit a value of β to experimental information on the wavelength of maximum absorption in the ultraviolet spectra of four polyenes. In this project, we shall predict the wavelength of the absorption maxima of the same four polyenes using the calculated difference (in units of electron volts) between the LUMO and HOMO of these four molecules. Bear in mind that this is not an *ab initio* calculation of wavelengths of maximum absorption, because empirically fitted parameters, $\beta, y_{11}, y_{12}, \ldots$ exist

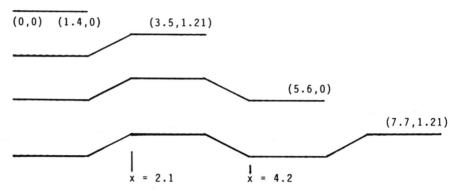

FIGURE 9-1 Approximate geometries for the alternant hydrocarbons ethene through octatetraene.

within the program or are calculated from empirical data by it. True *ab initio* calculations are yet to come.

Procedure Carry out SCF calculations for ethene, 1,3-butadiene, 1,3,5-hexatriene, and 1,3,5,7-octatetraene using program SCF. The program prints a series of prompts. You will need to designate the number of molecules to be run in any series, for example, by answering NMOLS? with 001. You will need to tell how many molecular orbitals will be calculated and how many are filled, for example, 004002 for 1,3-butadiene. You will need to specify the geometry by giving the x coordinates of all atoms in the molecule followed by the y coordinates. The molecule is assumed to be planar; hence, the z coordinates are all 0. Unformatted inputs are separated by commas, e.g., 0,1.4,2.1,3.5. Following this, the atoms are numbered in the same order as the coordinates were entered, obviously, 1,2,3,4 in this case. At this point, the coordinates are automatically printed out as an input check. The number of derivatives is asked, NDER?. For now, we are only interested in the parent compound; enter 0001. After more output, you are asked for OPTIONS? For now, choose 0000. Designators like I3 are input formats (Chirlian, 1981). I3 signifies an integer in the rightmost position in a field of three digits, e.g., 001.

Sketch the molecules on graph paper to help in determining the atomic coordinates. This is the first use of molecular geometry, a property that will become increasingly important as we go on. At this stage, the geometries are approximate; the difference, for example, between *cis* and *trans* isomers is ignored.

Determine the SCF energy difference between the LUMO and HOMO for these molecules. Program output is in electron volts. Convert to cm^{-1}. Predict the most intense ultraviolet absorption peak from the calculated energy separation. Plot the predicted wavelengths of maximum absorption against the number of double bonds. Plot the experimental values on the same graph paper. They can be found in Computer Project 7-2. Comment on the agreement between predicted and obtained wavelength maxima, or the lack of it.

COMPUTER PROJECT 9-2| *SCF Dipole Moments*

Unlike Huckel programs, SCF programs require an input geometry. Because charge densities are calculated, it is a simple matter to combine the atomic coordinates with the SCF charge densities (Chapter 8) to obtain a dipole moment. The charge densities are progressively refined by recalculating the matrix elements during the SCF calculation. In particular, alternation of long single and short double bonds is taken into account (partially) by the calculation. Hence, we expect better agreement between the calculated dipole moment and experiment than Huckel calculations gave.

The implication of self-consistency is that the calculation is iterated until calculated properties (strictly, one selected property) remain constant on renewed iteration. Program SCF, however, stops calculating new matrix elements after 10 iterations. The assumption is that self-consistency will have been reached by then.

We shall now investigate the OPTIONS? switch that was set to 0000 in Computer Project 9-1. At the prompt OPTIONS?, one has the choices 0, 1, 2, or 3 in I4 format. Taking them in reverse order, 0003 permits one to input a value for nuclear charge. It will not be used here, nor will 0002, which permits one to modify the repulsion integrals y_i. An option input of 0001 at this point can be used for heteroatoms by modification of the parent molecule matrix elements in a way that is almost the same as the heteroatom inclusion in HMO. We shall use this option. After all changes have been entered, the final option is 0000, which causes the calculation to be carried out. In Project 9-1, this was the only option entry; hence, the parent molecule itself was run.

Procedure

(a) Calculate the dipole moment μ of methylenecyclopropene using the geometry given in Chapter 8. Is the result closer to the (interpolated) experimental value of 1–2 D than the Huckel calculation?

(b) Investigate the influence of geometry on the dipole moment by "stretching" and "compressing" the methylene bond, that is, calculate μ at a methylene bond length of 1.0–1.6 Å at intervals of 0.1 Å. Leave the rest of the geometry the same as it is in part (a). What is the change in μ

in debyes per angstrom? There is a discontinuity at 1.5 Å. Why? Can you "fix" this problem to obtain reasonable values at 1.5 and 1.6 Å?

(c) Calculate the dipole moment of cyclopropenone using the OP-TIONS input to change the 1,1 matrix element. Use the results to infer the direction of the dipole moment toward or away from the methylene group.

COMPUTER PROJECT 9-3 | *Configuration Interaction*

Carry out SCF calculations for ethene, 1,3-butadiene, 1,3,5-hexatriene, and 1,3,5,7-octatetraene with full configuration interaction using program PPP. Program PPP has a far more elaborate input than any program we have yet run; thus, a different input strategy is used. A permanent, uniquely named input file is used. It can be run any number of times and can be copied or modified. If an error is made, the entire file need not be laboriously reentered. EDLIN is used to correct the mistake and the file is rerun. Modifications can be made to a copy of a successful input file to minimize the work of creating a new input file. In this way, a library of input files can be accumulated. New problems that are similar to old problems can often be solved by making a few modifications to an existing input file, thus saving the work of writing out a new file from the beginning.

In addition, the output of the run is stored in its own file. The output file can be printed once or repeatedly. It can be renamed for storage in an archive. The computational chemist is often interested in a special group of related molecules. The input/output (I/O) strategy just described permits the researcher to build an archive that is unique to his or her research field and may, with time, become the most complete archive in existence for that particular class of compounds.

Some version of this I/O strategy is used for all advanced programs to be discussed in the remainder of this book.

Procedure The most intense ultraviolet or visible absorption peaks are given in the table for each designated compound in Computer Project 7-2. Carry out SCF calculations for ethene, 1,3-butadiene, 1,3,5-hexatriene, and 1,3,5,7-octatetraene with full configuration interaction using program PPP. Use the input file above for ethene. It can be renamed CP3.DAT or it can be copied to CP3.DAT. After a successful run, the entire output file will be stored under the program name CP3.RES. Begin an archive by copying CP3.RES to ETHENE.ARC or some other unique file name. Copy the ethene file above to BUT.DAT and modify it, using EDLIN, to obtain the input file for the butadiene model. Copy BUT.DAT to CP3.DAT. (The old file there will be lost.) Run PPP again. Put CP3.RES into the archive under a unique file name. Continue for the hexatriene model and the octatetraene model.

The calculated absorption peaks are given in the output file CP3.RES, under the heading TRANSITIONS AFTER CI. Tabulate predicted absorption maxima for the four polyenes. Plot the predicted wavelengths of maximum absorption against the number of double bonds. Plot four curves on the same graph paper: HMO and experimental (from Computer Project 7-2), SCF (from Computer Project 9-1), and SCF–CI. Comment on their agreement or the lack thereof.

PROBLEMS | *Chapter* 9

1 Calculate the integrals H_{12}, H_{21}, and H_{22} that go with the integral calculated in Exercise 9-1.

2 Calculate the integrals S_{12}, S_{21}, and S_{22} that go with the integral calculated in Exercise 9-2.

3 Express the determinant

$$\begin{vmatrix} 2 & 3 \\ 4 & 5 \end{vmatrix}$$

as a single scalar. Exchange the columns and express the determinant as a single scalar again. What happens to the sign?

4 Given the linear combination

$$\psi = c_1 x (1 - x)^2 + c_2 x^2 (1 - x)$$

analogous to Eq. (9-4), find H_{11} by the method used in Exercise 9-1.

5 Given the linear combination in Problem 4, compute H_{12} and H_{22}.

6 Fill out the H matrix for the linear combination in Problem 4.

7 Determine S_{11}, S_{12}, and S_{22} for the linear combination in Problem 4.

8 Fill out the secular determinant for the linear combination in Problem 4.

9 Solving by the method of Computer Project 1-3, find both roots of the secular equation

$$(28 - 4x)(112 - 4x) - (7 - 3x)^2 = 0$$

which arises from the expansion of the determinant obtained in Problem 8.

10 From the lower root of the equation solved in Problem 8, determine the eigenvalue (energy) for the linear combination in Problem 4. This eigenvalue is analogous to the eigenvalue in Eq. (9-15).

REFERENCES

Anderson, J. M., 1966. *Mathematics for Quantum Chemistry*. Benjamin, New York, Section 3.2.

Chirlian, P. M., 1981. *Microsoft FORTRAN*. Dilithium Press, Beaverton, Oregon.

Greenwood, H. H., 1972. *Computing Methods in Quantum Organic Chemistry*. Wiley Interscience, New York.

Hinchliffe, A., 1988. *Computational Quantum Chemistry*. Wiley, New York.

Mc Quarrie, D. A., 1983. *Quantum Chemistry*. University Science Books, Mill Valley, CA.

Pariser, R. and Parr, R. G., 1953. *J. Chem. Phys.* **21**, 466, 767.

Pople, J. A., 1953. *Trans. Faraday Soc.*, **49**, 1375.

10 | *Molecular Mechanics*

A quite different approach to molecular energy and structure determination from the one we have taken so far is to regard the geometry of a molecule as the result of classical mechanics operating, through the medium of chemical bonds, on the connected masses of the atoms that constitute the molecule. This approach is called molecular mechanics (MM). Although different in concept, MM works hand-in-glove with the advanced molecular orbital methods like MNDO and *ab initio* computations, which will be described in Chapters 12 and 13.

The greatest limitation to molecular orbital calculations at present is the amount of computer time required to do them. This is, of course, a much more severe limitation on microcomputers than it is on mainframe computers. The problem can be diminished by using an accurate input geometry, just as any iterative procedure can be speeded up by a good initial guess at the answer.

Computation time is much less for MM calculations than it is for advanced MO methods. Therefore, a strategy often seen in contemporary computational chemistry is (i) calculate the energy and geometry of the target molecule by MM; (ii) if one has enough information to solve the problem, stop; (iii) if not, use the MM geometry to construct an input file for a semiempirical MO calculation (chapter 13), an *ab initio* computation (Chapter 12), or both.

Many molecules are too large to encourage *ab initio* or semiempirical MO calculations on existing computers, even with a good starting geometry. For this reason MM calculations as complete computations, not as a means to an end, are important, especially in biochemistry, molecular biology, pharmacology, and pharmaceutical chemistry. Research-level MM calculations are carried out (Rogers, 1988; Scott et al., 1988) using programs that are readily available, run on an off-the-shelf microcomputer with a math coprocessor. This chapter shows how to create and modify input data files and gives results for projects on propene and 1-butene.

HARMONIC MOTION

Imagine two stationary objects connected by a spring. One knows, of course, the distance between them; it is the length of the spring. Suppose the objects are in motion along the line connecting them. Without more information, one does not know the exact distance between them at any moment, but the average or equilibrium distance between them over a period of time is still the length of the spring when it is neither stretched nor compressed. (For an ideal spring, the average and equilibrium distances are the same.) If enough is known about the mechanical characteristics of the spring, it is possible to calculate the potential energy of the system at any separation of the objects. For any nonequilibrium distance, the potential energy will always be higher than it is at the equilibrium separation. Thus, one way to determine the equilibrium length of a spring within a system of springs is to calculate the energy of the system for many arbitrary configurations of all the springs and throw out all but the lowest-energy solution. If enough calculations are done at sufficiently fine discrimination, the equilibrium separation, hence the resting length of the spring, can be approximated to any desired accuracy.

In particular, Hooke's law gives

$$V = \frac{k}{2}(r - r_0)^2 + \text{const.} \qquad (10\text{-}1)$$

for the potential energy V of a single particle connected to a wall by a spring with a force constant k (Fowles, 1962). The displacement r from an equilibrium position r_0 at any time is

$$r - r_0 = \left(\frac{2V}{k}\right)^{1/2} + \text{const.} \qquad (10\text{-}2)$$

When the term on the right is at a minimum, $r = r_0$, the term on the left is 0 and the system is at its potential energy minimum.

Problems of two masses connected by a spring are worked in essentially the same way.

If there are many objects connected by many springs, there will be an overall equilibrium configuration that leads to an overall potential energy minimum, but this need not and, in general, will not be at the rest lengths of all or any of the springs. The equilibrium configuration will be a compromise of all forces acting on all objects.

One now seeks to find a matrix of interparticle distances **R** by inverting the matrix of force constants **K** connecting the masses:

$$\mathbf{R} = (2V\mathbf{K}^{-1})^{1/2} \qquad (10\text{-}3)$$

for the simple case of harmonic oscillators. This can be done for many interparticle distance matrices $\mathbf{R}_i$, and the one yielding the lowest potential energy matrix **V** can be selected as the arrangement of masses that has the best compromise of stretches, compressions, bends, etc., of the connecting springs.

If one starts with an initial guess $\mathbf{R}_0$ as the matrix of all interparticle distances, after many displacements of the masses, the cumulative displacement of each mass that minimizes the potential energy constitutes a displacement matrix $(\mathbf{R} - \mathbf{R}_0)$. Knowing the x, y, and z coordinates of the initial guess, one has the x, y, and z coordinates of all masses after the displacements that brought about energy minimization.

MOLECULAR MECHANICS

If the objects in the last few paragraphs are atoms in a molecule and the springs are chemical interactions in the most general sense, including chemical bonds and repulsive and attractive interactions of all kinds, then, at the potential energy minimum, the atoms will be in a configuration that is their most stable molecular structure. Because the energy was minimized to obtain the overall equilibrium structure, both energy and structure emerge from the same procedure.

In molecular mechanics, physical springs of the classical analog are replaced by mathematical functions that describe the various interactions within the molecule. It turns out that Hooke's law is a rather good approximation to the stretching mode of a chemical bond. Torsional forces and van der Waals attractions and repulsions must also be included. Force constants of the springs are represented by a collection of mathematical parameters. Collectively, the parameters are called a *force field*. They are used, in conjunction with classical equations of motion, in determining the overall energy of any configuration of atoms, i.e., any molecular structure.

Huge numbers of random guesses as to the molecular structure are, of course, not made. Rather, a single plausible structure is guessed and iterative changes are made in that structure in a methodical way so as to approach the minimum energy most efficiently, in the way that an intelligent blind man would find his way down hill.

The potential energy of the system of n atoms in a molecule can be expanded in a Taylor series over the degrees of freedom for each atom (Burkert and Allinger, 1982). Within the Hooke's law approximation, most terms drop out, leaving

$$V = \tfrac{1}{2} \sum_{i,j=1}^{3n} \left(\partial^2 V / \partial x_i\, \partial x_j \right)_{x_0} \Delta x_i \, \Delta x_j \tag{10-4}$$

The terms Δx_i and Δx_j are deformation terms, i.e., displacements away from an equilibrium coordinate x_0.

The second derivative in Eq. (10-4) is a force

$$V = \tfrac{1}{2} \sum_{i,j=1}^{3n} k_{ij} \, \Delta x_i \, \Delta x_j \qquad (10\text{-}5)$$

All force constants k_{ij} in Eq. (10-5) constitute a matrix **K**. If the matrix is diagonal, the system is a collection of uncoupled oscillators, analogous to a set of simultaneous equations in which each equation has only one nonzero independent variable. This is not an interesting system for us. Molecules are collections of atoms that oscillate in a mutually dependent way; they are coupled. Diagonalization of the K matrix yields the eigenvalues of the system, which lead to the energy contribution from each mode of vibrational motion. If all modes of motion, vibrational, rotational, etc., are included, the result is a strain energy that can be added to the strain-free energy of a molecule to obtain its actual energy. Strain-free energies are calculated from some collection of ideal bond energies obtained from simple molecules that are presumed to be strain-free (Computer Projects 6-2 and 6-3). In rough usage, energy and enthalpy are taken as synonymous.

Using an MM program, then, requires an initial geometry of the target molecule, the INPUT file. Operation of the program involves many changes in the coordinates of each atom in the molecule, solution of the potential energy equations, and a methodical (iterative) approach to the overall energy minimum. Iterations continue until further small changes in the positions of the atoms no longer change the energy. Equations for V and its first and second derivatives (slope and curvature) are all rather simple, partly accounting for the speed of molecular mechanical calculations relative to advanced molecular orbital methods.

THE PROGRAM

The MM program [QCMPE 004 (Chong, 1985)] as received from QCPE is on five disks, numbered ZERO through FOUR. The first three are in executable code and the remaining two are in FORTRAN source code. The first three disks are all you need to get started unless you wish to modify MM (not recommended). Copy the masters and, at the system prompt A > put copies of disks ZERO and ONE into drives A and B. The command STEP0 causes initialization and clears certain working files, in particular, the UNIT7 and OUTPUT2 files. When this is done, the directory is printed to the CRT screen. IF you try to TYPE files UNIT7 or OUTPUT2, you will find that they are empty.

At this point, the command STEP1 is entered, the input data are read, and modifications, if any, are made on the input data as required by the input format (see below). For the demonstration input file, included with the software package, this takes about 3 min on an 8086-based machine at 8 MHz. A Stop-Program terminated message to the CRT screen tells you that the data have been read. At this point, replace disk ONE with a disk TWO in the B drive and enter the command STEP2. This run takes about 45 min for the demonstration input file. After a successful run, a Stop-Program terminated message is printed to the CRT screen and the system prompt A > reappears.

Run the directory again. You will note that several files that had only 1 byte of information (a place holder for an empty file) now have hundreds or thousands of bytes of information. The OUTPUT1 file contains the modified input and the initial steric energy, which is high because geometric optimization had not yet taken place when OUTPUT1 was written. The output of the minimization procedure is obtained by using the command TYPE OUTPUT2. This displays the results of the molecular mechanics calculation on the CRT screen or printer. The output of the demonstration file is more lengthy than we wish to describe here; we defer that discussion to the output of our first MM run from scratch. Suffice it to say that the output should have a block describing some number of iterations followed by a block giving optimized atomic coordinates, distances, and bond angles and, finally, a block containing enthalpy information. If your demonstration compound is cyclotetradecane (the demonstration file for QCMPE 004), the enthalpy of formation will be about -69 kcal mol^{-1}.

Creating an Input File

A good deal of labor can be saved by building a molecular mechanics input file in a logical way from simpler to more complex molecules. In particular, there are attachment options in the MM program that permit us to build molecules group-by-group or to enter just the carbon skeleton and to attach hydrogens, methyl groups, and phenyl groups as desired.

A (not very good) starting geometry for the ethene molecule is given in Fig. 10-1. The two carbon atoms have been placed on the x axis, equidistant from the origin, 1.4 Å apart. The hydrogen atoms have been symmetrically placed 0.5 Å above and below the x axis at a horizontal distance of 1.0 Å from the origin. Most people will notice that the molecule looks too long and thin for ethene; the geminal hydrogens are

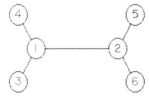

FIGURE 10-1 Molecular mechanics starting geometry for ethene.

too close together and the carbons are slightly too far apart. Let us number the carbon atoms 1 left, 2 right and the hydrogens 3, 4, 5, and 6 counting clockwise from the lower left.

Make a backup of the demonstration input file. The old input file serves as a pattern for constructing new files. A pattern is necessary because input formats are very strict and it is much easier to strike over old input than to write new files from scratch. For your first file, delete all lines except the first seven in the old INPUT file. Lines are called up for alteration by entering the line number **n**; changes are made by moving the cursor and striking over; a line or lines before **n** are inserted using the edit command **n***i*. A ruined file can be erased; you can always recopy the backup. Exit the editor by typing *e* at the edit prompt *∗*.

Later, you will wish to edit UNIT7. Copying the edited version of UNIT7 to INPUT permits you to run the edited file (see below).

The Ethene Input File

The input file for ethene is given in Fig. 10-2. The largest block of information occupies the lower half of the figure and consists of 18 floating-point numbers in six groups of three, each group followed by an integer. The floating-point numbers are the x, y, and z coordinates (in that order) for the six atoms in ethene. The integers give the identity of that atom. For example, the first group of three numbers is a -0.7 followed by two entries of 0.0, indicating that as a first approximation we have taken one atom to be on the x axis (y and z coordinates equal 0), 0.7 Å to the left of the origin. The coordinates are followed by the integer 2, which designates the atom in question as an sp^2 carbon. In this chapter, we shall use only the integers 1 (signifying an sp^3 carbon), 2 (an sp^2 carbon), and 5 (a hydrogen). The rest of the coordinates correspond to the trial geometry in Fig. 10-1 for ethene. The z coordinates are all taken as zero for convenience, that is, the molecule is placed in the x-y plane.

The top half of the input file is devoted to describing the connection pattern of the atoms, along with some control information. The first line contains the alphabetic information "ethene" followed by a 6, which gives

FIGURE 10-2 Possible input file for ethene. Many other input files will arrive at the correct result.

	ethene									6 1		
1				4						1		
1	2											
1	3	1	4	2	5	2	6					
-.7	0.0	0.0			2		.7	0.0	0.0		2	
-1.	-.5	0.0			5		-1.	.5	0.0		5	
1.	.5	0.0			5		1.	-.5	0.0		5	

the number of atom x, y, z coordinate groups to be read in, and a 1, which specifies the amount of calculated information to be printed out. Line 2 stipulates that there will be 1 "connected atom list," 4 "attached atoms," and the 1 immediately below the 6 in line 1 stipulates that heat of formation calculations will be included in the printout. Line 3 is the connected atom list referred to; it says that atom 1 is connected to atom 2 (the two carbon atoms). Line 4 is the attached atom list; it says that atom 3 is attached to 1, atom 4 is attached to 1, atom 5 is attached to 2, and atom 6 is attached to 2. These are, of course, the hydrogens. Input formats are very strict; all control information must be in the correct column.

The run time for this small molecule is about 5 min on an 8086 microprocessor at 8 MHz.

The Output File

If only the optimized geometry is desired, TYPE UNIT7. The UNIT7 file for ethene is given in Fig. 10-3. During optimization, the x component of the carbon separation distance has contracted to 1.33 Å and the hydrogens are now separated by 1.88 Å in the y direction. These separations (components) are not exactly the atom-to-atom separations because the molecule has rotated a little during minimization. It also has moved a little to the right (translation in the x direction).

A more complete file is obtained by typing OUTPUT2. Note that Shift PRINT dumps the screen to the printer. Pressing the PRINT (or CTRL PRINT) key before entering the TYPE OUTPUT2 command causes the output file to be routed to the CRT and the printer at the same time. The PRINT key toggles; that is, pressing it a second time cuts the printer out while the CRT remains active. The HOLD key can be used to stop scrolling of the CRT display when you wish to inspect a part of the output. The HOLD key also toggles.

The entire output file will not be reproduced here [see Clark (1985) for extended printouts]; however, some of its features will be discussed.

Ethene goes through 15 iterations on the first cycle. The average movement of atoms converges on zero and the total steric energy converges on 0.42 kcal mol $^{-1}$. In cycle 2, UNIT7 is written (lines designated PUNCHED). The final coordinates are printed out along with bond

FIGURE 10-3 Optimized geometry for ethene.

ethene						6 0		
-.63455	-.22576	.00000	2	.68457	.00508	.00000	2	2
-1.03247	-1.25370	.00000	5	-1.35790	.60597	.00000	5	4
1.08248	1.03302	.00000	5	1.40791	-.82665	.00000	5	6

lengths of 1.34 Å for C=C and 1.10 for C—H. Bond angles are 118–121, i.e., about 120°. Dihedral angles are for the molecule viewed end-on as in a Newman projection. Thus, hydrogen atoms 4 and 5 exactly eclipse one another as do 3 and 6. Hydrogens 3 and 5 and 4 and 6 are exactly anti (180°). Bond enthalpies are summed with the total steric energy in the last block of output to yield 12.84 kcal mol^{-1} for the enthalpy of formation [experimental value: 12.45 (Cox and Pilcher, 1970; Pedley, Naylor, and Kirby, 1986)]. The total steric energy is the sum of energy increments due to intramolecular repulsive forces and to any bending or stretching that has to take place in order to reach an optimum geometry. In ethylene, there is essentially no steric interference, so the steric energy is essentially zero.

Modifying the File

Using EDLIN B:UNIT7, with the MM2 ZERO disk in drive B, one can change line 1 of the UNIT7 file produced during ethene minimization to read propene 6 1 as in Fig. 10-4. A line 1 4 1 1 can be inserted as line 2, and the connected and attached atoms lists can be reinserted as lines 3 and 4. One now has an input data file with (aside from the alphabetic change to "propene") two important differences from Fig. 10-2. First, the coordinates are no longer approximate, but have been optimized and are exact *for ethene* (within the limits of the force field). Second, an extra 1 appears as the third entry in line 2. This number calls for one addition or substitution line. We can provide this by inserting line 8 at the end of the file: 7 6 2 1 as in Fig. 10-4. These numbers signify that a methyl group will be substituted (7) for hydrogen atom 6, which is attached to carbon 2, which, in turn, is attached to carbon 1. The numbers are needed to designate, unequivocally, the atom you want substituted. Substitution (replacement) by a methyl group is one option (7) among several that include (1) addition of one hydrogen to an sp^3 carbon, (2) addition of two hydrogens to an sp^3 carbon, (4) addition of one hydrogen to an sp^2 carbon, and (9) substitution of a phenyl group.

When modification of the UNIT7 file has been completed and it is as shown in Fig. 10-4, we have an input file for the propene molecule. The

FIGURE 10-4 Optimized geometry of ethene modified to produce an input file for propene.

	propene							6 1		
1			4		1			1		
1	2									
1	3	1	4	2	5	2	6			
-.63455	-.22576	.00000	2			.68457	.00508	.00000	2
-1.03247	-1.25370	.00000	5			-1.35790	.60597	.00000	5
1.08248	1.03302	.00000	5			1.40791	-.82665	.00000	5
7	6	2	1						

	propene					9 0		
-.52192	-.11971	-.61799	2		.69129	-.00118	-.06045	2	2
1.67177	-1.13605	.06378	1		-.87478	-1.08026	-1.02587	5	4
-1.20370	.74385	-.68448	5		1.00683	.97929	.33690	5	6
2.61557	-.88613	-.47223	5		1.27614	-2.08571	-.36308	5	8
1.91363	-1.31643	1.13593	5					LAST

FIGURE 10-5 Optimized geometry for propene.

COPY UNIT7 INPUT command copies the modified file to the input of MM2. Running the new file generates an enthalpy of formation for propene of 4.91 kcal mol^{-1} [experimental value: 4.88 (Cox and Pilcher, 1970; Pedley, Naylor, and Kirby, 1986)] and a set of coordinates that is in the new UNIT7 (Fig. 10-5). Plotting the *x-y*, *x-z*, or *y-z* coordinates, taken from either the UNIT7 file or the OUTPUT2 file, in two dimensions gives the three orthogonal projections of the propene molecule (Fig. 10-6).

Clearly, the process just described can be repeated so as to generate MM results for an indefinite number of alkenes. Methyl substitution for hydrogens 4 and 5 of the UNIT7 file for propene leads to the correct relative stabilities for the *cis* and *trans* isomers of 2-butene (Computer Project 10-2). Methyl substitution for the 7, 8, and 9 hydrogens of propene leads to conformers of 1-butene (Computer Project 10-3). Starting with an input geometry for ethane enables one to build up the alkane series. As one builds toward larger molecules, there are very many choices that can be made, leading to different isomers and conformations.

Cyclopentane

Additions and substitutions do not have to be made one at a time. Calculation of the molecular structure and energy of cyclopentane provides a simple example of the use of continuation lines to build a starting structure. The idea is to write an input file for the carbon skeleton only,

FIGURE 10-6 Stick diagram of the optimized geometry of propene. Dotted and bold lines for receding and protruding hydrogen atoms have been added to the original diagram.

then modify it to the input file for cyclopentane by adding hydrogens to the skeleton. Once a starting structure has been run successfully, many derivatives can be made by modifying the input file using substitution options in ways already shown, (e.g., substituting a methyl group for a hydrogen atom).

First, lay out a pentagon on graph paper. Estimate the x-y coordinates for the carbon atoms and let all the z coordinates be zero. The estimated coordinates need not be entered with great accuracy; they will relax to the correct values as minimization proceeds. There are a few reasonable choices as to where the pentagon can be situated relative to the origin. We chose to let the base of the pentagon be the x axis with two of the five carbons, arbitrarily numbered 1 and 2, equidistant from the origin. Carbon–carbon bond lengths are taken to be 1.4 Å. In program MM2, all bond lengths are in angstroms and all energies are in kilocalories per mole.

The x, y number pairs {0.7, 0.0; −0.7, 0.0; −1.1, 1.0; 0.0, 2.0; 1.1, 1.0} for the coordinates of the carbon atoms restricted to the x-y plane give a roughly pentagonal starting geometry shown in Fig. 10-7. Leading zeros

FIGURE 10-7 Possible input file for cyclopentane. The ball-and-stick diagram is for input geometry and has not been optimized.

```
                cyclopentane                                    5 1
  1                                        5                     1
  1   2    3    4    5    1
   .7      0.0       0.0        1      -.7      0.0      0.0      1
 -1.1      1.0       0.0        1      0.0      2.0      0.0      1
  1.1      1.0       0.0        1
  2   1    2    5
  2   2    1    3
  2   3    2    4
  2   4    3    5
  2   5    4    1
```

(a)

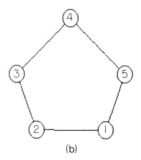

(b)

can be used with decimals, e.g., -0.7; they are good form and make reading data easier but are not strictly necessary. Carbon atoms are numbered $1, 2, 3, 4, 5$ clockwise, starting at the lower right. Other numbering systems work just as well. Including the z coordinates, all of which are 0 gives 15 coordinates, three for each atom, which are entered in two columns and constitute the middle block of information in the input data file, Fig. 10-7. After each triplet of coordinates, is an atom type designator. In Fig. 10-7, each atom type designator is 1, indicating an sp^3 carbon.

The entire input file consists of three blocks of information: a control block, a coordinate block (already described), and a continuation block. The lowest block of numerical information in Fig. 10-7, five lines of evenly spaced single-digit numbers, is the continuation block. These five lines cause the input file, which is for the five-carbon skeleton only, to be modified to the input file for cyclopentane by adding two hydrogens to each carbon atom in the original file.

The structure of a continuation line is such that the first digit causes the action; in this case, the digit 2 causes addition of two hydrogens to an sp^3 carbon. Another action digit that we shall use is 4 for attaching one hydrogen to an sp^2 carbon atom. The second number in a continuation line designates the atom to which hydrogen is to be attached. The third and fourth numbers specify atoms attached to the atom designated by the second digit. Thus, the continuation line 2 1 2 5 conveys the information *attach two hydrogens to* sp^3 *carbon atom* 1 *in the tetrahedral positions opposite connected carbon atoms* 2 *and* 5. The remaining four continuation lines complete attachment of all 10 hydrogens in cyclopentane.

The first line in the control block has an alphabetic identifier followed by the number of atoms in the input structure, *not counting those added by the continuation lines*. The number of atoms in the skeleton is 5. The second line in this control block contains the digits 1 5 1 indicating, respectively, that there is 1 connected atom list, there are 5 continuation lines, and that the enthalpy of formation will be calculated. Omitting the final 1 yields a normal geometric minimization but no enthalpy of formation. (Blank is read as 0, which usually signifies negative.) Entries that are either 0 (off) or 1 (on) are often referred to as *switches*. The third line in the control block is the connected atom list. Note that atom 1 is connected to atoms 2 and 5 in the cyclopentane skeleton.

The Program Run

STEP0 clears internal memory locations to prepare for the calculation. STEP1 performs the attachments indicated by the continuation lines and generates a "normal" input file, i.e., one with no continuation lines and all hydrogen atoms in place in the sense that the correct *starting* bond angles and lengths have been used to specify coordinates for the hydrogen atoms attached to the planar ring of carbon atoms. The starting file can be seen by executing TYPE UNIT7 after STEP1 but before STEP2. OUTPUT1 also contains the starting geometry. None of these initial coordinates is correct because the aggregation of atoms has not been relaxed into its low-energy conformation by the minimization procedure. UNIT7 before minimization is shown as the top half of Fig. 10-8.

The actual minimization is carried out using the command STEP2. After STEP2, the command TYPE UNIT7 produces a file of the same form as it did following STEP1 but with some of the control information stripped off and different coordinates. These are the equilibrium coordinates after minimization, and together they constitute, within the limits of the force field, the best molecular structure for that input file. OUTPUT2 gives the final geometry along with many geometric parameters, such as bond angles, bond lengths, and dihedral angles.

The structure is that of whichever conformer is geometrically nearest the starting structure. UNIT7, after minimization, is shown as the second half of Fig. 10-8. Two views of cyclopentane are shown in Fig. 10-9.

FIGURE 10-8 (a) Working input file (without addition lines) for cyclopentene and (b) output file for cyclopentene. The energy has been minimized and the geometry is optimized.

		cyclopentane										15	0	0	
1				0	10		0								
1	2	3	4	5	1										
1	6	1	7	2	8	2	9	3	10	3	11	4	12	4	13
5	14	5	15												
.70000	.00000	.00000	1		-.70000	.00000	.00000	1							
-1.10000	1.00000	.00000	1		.00000	2.00000	.00000	1							
1.10000	1.00000	.00000	1		1.15304	-.45304	-.88183	5							
1.15304	-.45304	.88183	5		-1.15304	-.45304	.88183	5							
-1.15304	-.45304	-.88183	5		-1.74069	1.00000	.88183	5							
-1.74069	1.00000	-.88183	5		.00000	2.64069	.88183	5							
.00000	2.64069	-.88183	5		1.74069	1.00000	.88183	5							
1.74069	1.00000	-.88183	5												
		cyclopentane										15	0		
.72357	-.26118	.40481	1		-.74468	-.32980	-.03980	1	2						
-1.25557	1.12287	-.00489	1		-.02216	2.00658	.25102	1	4						
1.17615	1.11469	-.09914	1		1.34111	-1.09630	-.00365	5	6						
.78395	-.28940	1.51932	5		-1.34564	-1.01259	.60687	5	8						
-.79187	-.71998	-1.08498	5		-2.02037	1.26992	.79471	5	10						
-1.73094	1.38717	-.97980	5		.02929	2.27524	1.33365	5	12						
-.03666	2.95375	-.33905	5		2.12655	1.45866	.37419	5	14						
1.31812	1.08493	-1.20609	5						LAST						

(a)

(b)   **FIGURE 10-9** Stick diagrams for cyclopentane.

False Minima

In trying alternate ways of building up input files, one is bound to discover paths that lead to different enthalpies of formation for the same molecule. For example, one might happen upon an input geometry for propene that leads to a normal minimization, but results in a calculated enthalpy of formation that is about 2 kcal mol^{-1} above the thermodynamic value. Does this mean that the arbitrary selection of starting geometry determines the outcome of an MM calculation? Yes and no.

There is no known computational method for finding the absolute energy minimum in a potential field. In general, MM calculations seek the extremum closest to the starting geometry. Not all extrema are minima. Molecular geometries may come to rest at a local minimum or (rarely) a saddle point on the potential energy surface. Different conformers satisfy the mathematical requirements for a local minimum but are not necessarily at the absolute or global minimum. Butane, with its *anti* and *gauche* forms constitutes a classic example and is treated in many elementary texts (Morrison and Boyd, 1973). Butane has three minima, one of which, the *anti*, is the global minimum.

False minima are a pitfall for the unwary; a structure may be calculated that does not represent the most stable conformer. If so, the calculated enthalpy of formation will always be higher than the thermodynamic enthalpy of formation. Conversely, false minima can be just the

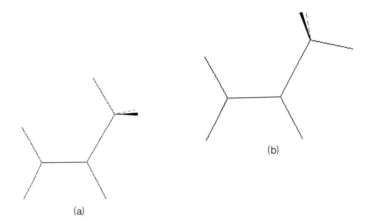

FIGURE 10-10 Stick diagrams for the two lowest-energy conformers of propene. (a) The global energy minimum (4.91 kcal mol^{-1}). (b) The next higher-energy conformer ($\Delta H_f = 6.99$ kcal mol^{-1}).

thing one wishes to find in a conformational study. Drawing the optimized geometries of low- and high-enthalpy forms of propene, for example, soon persuades us that the stable form has a methyl hydrogen that eclipses the double bond and the high-enthalpy form has two hydrogens in a staggered conformation to it (Fig. 10-10). This is contrary to the case for alkanes in which the staggered conformation is more stable than the eclipsed.

In the case of small molecules, one can determine the enthalpy of formation for all of the limited number of conformations that can exist. Depending on molecular rigidity and symmetry, the number of conformational choices may increase rapidly for large molecules. More complicated molecules present a challenge to the chemist's structural intuition because they often relax to local minima. A good structural analysis presumes that enough starting geometries have been tried (however many that may be), that no reasonable starting geometry has been missed, and that the lowest-energy conformer is indeed at the global minimum of energy. Whereas no mathematically rigorous proof exists that a conformer is at the global minimum, reasonable doubt can be made very small by judicious choice of starting geometries.

The opposite approach to judicious choice (Saunders, 1987) employs a program that perturbs the starting geometry in a random way, then allows it to relax. After many random "kicks" of this kind, the output files are examined for the lowest energy. Although time-consuming, the method has been shown to be effective in reproducing previously known global minima and increasing confidence in minima that are thought to be global. Saunders' method is probably too time-consuming for use with microcomputers at this writing, but in the next generation of microprocessors and at clock speeds attainable in the next few years, this drawback may disappear.

Dihedral Angles

Included in the information in OUTPUT2 for propene are 10 dihedral angles, each dihedral angle involving four atoms. The angles we are especially interested in involve carbon atoms 1, 2, and 3, which define a plane, and the methyl hydrogens, atoms 7, 8, and 9.

For ground-state propene (a), dihedral angles of 0, − 120, and 120° show an eclipsed conformation between carbon (1) and one of the methyl hydrogens (Fig. 10-10). These dihedral angles for the high-enthalpy conformer (b) are essentially 180, − 60, and 60°, indicating a methyl hydrogen anti to C(1). Deviations from exactly 0, 60, 120, and 180° in the dihedral angles are probably real, not artifacts of the calculation. Warping of alkenes as well as bending of normal bond angles is common.

The Programs

The programs discussed are QCMPE 004 and MMX, the Serena Software version of Allinger's MM2 in the 1985 modification (Appendix). These programs are modifications of QCPE 395. MMX performs pi-electron SCF calculations and has the option to drive dihedral angles. See the documentation of either program for details on these advanced options.

COMPUTER PROJECT 10-1 | The Enthalpy of Hydrogenation of Ethene

In this experiment, we shall run two molecular mechanics files: one for ethene and one for ethane. Each run gives a value of the enthalpy of formation of the target compound (along with a lot of other information). From these values, calculate the enthalpy of hydrogenation H_h for the reaction (Atkins, 1986)

$$CH_2 = CH_2 + H_2 \longrightarrow CH_3 - CH_3$$

Compare your answer with the experimental value -32.58 ± 0.05 kcal mol^{-1} (Cox and Pilcher, 1970; Pedley, Naylor, and Kirby, 1986). What is the percent discrepancy between the calculated value and the experimental value?

Procedure MM2 does not accept data from the keyboard as smaller programs do. The input data are in the form of a strictly formatted file that is unique to each molecule studied. After constructing the data file, it should be copied to a file called INPUT. During the program run, the computer reads whatever data are in file INPUT. When one wishes to study a different molecule, a new data file is made or a copy of an old data file is modified to reflect the differences between the old problem and the new problem (see the preceding section on input files). The new file can then be copied to INPUT. The old contents of INPUT will be lost when the new file is copied.

A different strategy is used by many programs. A prompt is presented on the CRT screen asking the name of the input file to be read. In this strategy, data will be read by answering the input prompt with the name of a file, e.g., ETHENE.DAT. Program QCMPE 004 uses the first strategy; MMX uses the second.

Following successful execution, you will obtain output from which you can follow the iterative minimization of the molecular energy in the force field by systematic variation of the atomic coordinates. If you are keeping an archive, copy INPUT to a unique file name after a successful run. Subsequent to this, structural information, which is not necessary in this project, will be printed out. At the end of the output, energies of various kinds are calculated, including the "heat" (strictly, enthalpy) of formation that we seek. When you have the enthalpies of formation of ethene and ethane, use Hess' law to obtain the enthalpy of hydrogenation. Remember that the enthalpy of formation of an element in the standard state is zero.

COMPUTER PROJECT 10-2| Stability of cis- and trans-2-Butene

Beginning with the UNIT7 file for propene, identify the hydrogens that will give *cis* and *trans* 2-butene when replaced by methyl groups. If the propene file has been created as described in this chapter, these will be hydrogen atoms 4 and 5. You may wish to use a graphics program (Chapter 11) to identify the hydrogen atoms we seek. Using a methyl substitution line, create two input files, one for the *cis* isomer and one for the *trans*. Minimize both structures. Determine the enthalpy of formation of each isomer. Which is more stable? Compare your results with the discussion of *cis–trans* isomerism as given in a typical elementary organic chemistry text.

COMPUTER PROJECT 10-3| Conformational Chemistry of 1-Butene

Substituting methyls for the hydrogen atoms on the sp^3 carbon atoms of propene and changing the input coordinates yields the MM solutions for several conformers of 1-butene, including a *dl* pair with C_1 symmetry and a conformation, having C_s symmetry, in which the methyl group eclipses the double bond. MM calculations show that these are the two lowest energy conformers.

Electron diffraction and microwave spectroscopic experiments show the eclipsed or *syn* rotamer to be present in a ratio of 17% of the total, the remaining 83% being the enantiomeric pair of *skew* rotamers (van Hemeluijk et al., 1980). Higher enthalpy forms make a negligible contribution to the rotameric mix. This leads to an enthalpy difference of 0.53 ± 0.42 kcal mol^{-1} between the two dominant conformers. MM calculations show the enantiomeric pair to be more stable than the *syn* rotamer by 0.49 kcal mol^{-1}. An *ab initio* calculation (Chapter 12) yields 0.60 kcal mol^{-1}.

Procedure Run MM calculations and draw a conformer that does not make a significant contribution to the conformational mix. Using Boltzmann's equation (Chapter 2), compute the percentage ratio (e.g., $17/83 = 20.5\%$ *syn*) of the conformational mixture of the two dominant conformers at the MM energy difference of 0.49 kcal mol^{-1}.

COMPUTER PROJECT 10-4 | Structure and Energy of Cyclopentene

Once cyclopentane has been run successfully, the input file should be backed up. One copy can then be modified for cyclopentene by changing the alphabetic designator, changing the atom type designator in the coordinate block from 1 to 2 for two carbon atoms, say atoms 1 and 2, and changing the continuation lines for carbon atoms 1 and 2 so that they have an action digit of 4 in place of 2. The new input file functions as the cyclopentane input file did except that one hydrogen atom each is attached to carbon atoms 1 and 2, which are now designated sp^2 rather than sp^3.

Molecular diagrams of cyclopentene in Fig. 10-11 show shortening of the double bond and partial eclipsing of hydrogen atoms. Some strain is relieved by relaxation to the "envelope" form with four carbons in a plane and one out of the plane. The envelope structure of cyclopentene can be seen in the projections—PCMODEL (Chapter 11) allows one to rotate the molecule to almost any angle—and in the dihedral angles.

Procedure Modify a copy of the cyclopentane file (to a unique file name) as previously described and run the modified file. The full printout of MM2 is far too bulky to reproduce here, but it contains a block of all

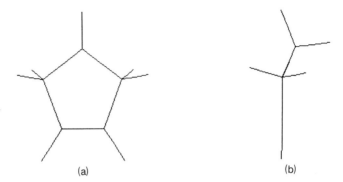

FIGURE 10-11 Stick diagrams for cyclopentene. The envelope conformation is evident in (b), which is the structure in (a) rotated 90° about its vertical axis.

dihedral angles in the molecule. Preceding the block of dihedral angles is a statement in the printout showing how they should be read and interpreted. The dihedral angle looking down the 1–2 bond (angle 3 2 1 5) is very nearly 0° because of the planar bonding of the sp^2 carbons 1 and 2. What is the dihedral angle of the 2 1 5 4 bond?

Partial eclipsing causes strain in the cyclopentene structure, which is calculated by MM2 to be 7.1 kcal mol^{-1}. Partial eclipsing is more severe in cyclopentane than it is in cyclopentene even though there is some relaxation of the cyclopentane molecule away from the eclipsed form. How much does strain energy increase on going from cyclopentene to cyclopentane? Propose an explanation on the basis of molecular crowding.

Cyclohexane, which is thought to be essentially unstrained due to release of crowding when it relaxes to its chair conformation, has an enthalpy of hydrogenation of -28.4 ± 0.2 kcal mol^{-1}. Predict the enthalpy of hydrogenation of cyclopentene on the basis of destabilization due to crowding. The experimentally observed value is -26.9 ± 0.2 kcal mol^{-1}.

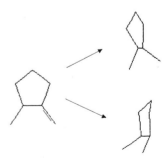

FIGURE 10-12 Possible mechanism for drogenation of 1,2-dimethylcyclopentene.

COMPUTER PROJECT 10-5| Enthalpy of Hydrogenation of 1,2-Dimethylcyclopentene

After a run on either cyclopentene or cyclopentane, the resulting UNIT7 file can be copied to a different name. This file can be modified by adding a methyl group or groups as shown in this chapter, to give the input file for a new molecule, say 1,2-dimethylcyclopentene. New control information must be written in and a designator must be given for two continuation lines. The lines themselves are 7 6 1 2 and 7 7 2 1, which substitute methyl groups for hydrogens on the two sp^2 carbon atoms. This new file can be copied to INPUT and run to yield the enthalpy of formation of 1,2-dimethylcyclopentene.

The same procedure can be used to carry out MM2 calculations on 1,2-dimethylcyclopentane, but now there are two ways of connecting the methyl groups, one *cis* and the other *trans*. Molecular graphics (Chapter 11) may be useful here to locate the hydrogens to be replaced by methyl groups. PCMODEL has a command that flags atoms by number.

When the *cis-* and *trans-* 1,2-dimethylcyclopentane files are minimized, they yield two values for the two possible reaction products. Gas chromatographic analysis of the hydrogenation product of 1,2-dimethyl-cyclopentene shows it to be 86% trans product and 14% cis (Allinger *et al.*, 1982). Calculate the weighted average enthalpy of formation of the hydrogenation product mixture. Use Hess' law to calculate the enthalpy of hydrogenation of 1,2-dimethylcyclopentene to the observed mixture of isomers. The experimental value is -22.5 ± 0.2 kcal mol^{-1}. Back calculation yields the enthalpy of hydrogenation of the reactant to each of the pure isomers. This result is unusual, not only because it is a thermochemical analysis of parallel hydrogenation reactions, but because the enthalpy of hydrogenation is one of the smallest ever measured for a hydrocarbon. The reactant molecule is stabilized by methyl groups in the 1 and 2 positions and the product, particularly the *cis* isomer, is destabilized by crowding. Both of these factors make hydrogenation less exothermic.

We find it interesting to speculate on why surface-catalyzed hydrogenation yields such a high percentage of trans isomer or, indeed, why it yields any trans isomer at all. One would, after all, expect cis addition across a planar double bond by hydrogen adsorbed on a flat surface. It is possible that direct hydrogenation of the tetrasubstituted double bond does not occur, but that preliminary migration of the double bond to the methylene intermediate does, after which hydrogenation of the methylene intermediate goes by a mechanism that is controlled by thermodynamic or other factors. Assuming the entropy difference between isomers to be zero ($\Delta G = \Delta H$), what is the ratio of isomers calculated from the molecular mechanics difference in enthalpy of formation of the isomers, $-RT \ln K = \Delta G$?

PROBLEMS | *Chapter 10*

1 Under what circumstances is the maximum speed numerically equal to the maximum excursion for a simple harmonic oscillator?

2 Write an expression for the kinetic energy T in a system of two masses connected by three springs. The central spring is called a coupling spring. Write an expression for the potential energy V for this system. Write an expression for the total energy E. Note the similarity between this expression and the electronic hamiltonian for helium (Chapter 12).

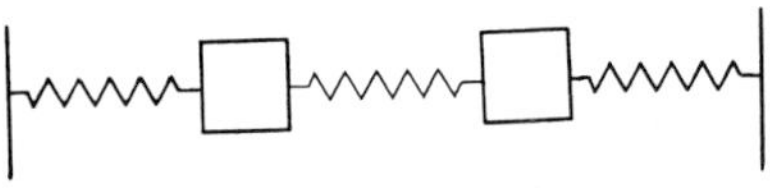

FIGURE P10-2

3 Plot the potential energy V for a spring with a force constant of 10 N m^{-1} over several displacements from -0.20 m through 0 to $+0.20$ m. What are the units of V?

4 A spring stretches 0.200 m when it supports a mass of 0.250 kg. (a) What is the force constant k of the spring?
If the mass is pulled 0.0500 m below its equilibrium point and released:
(b) What is the frequency of oscillation of the mass? What is its period?
(c) What is the potential energy of the mass at the extremes of its excursion? (d) What is the maximum speed of the mass?

5 Write and run the input file for methane "from scratch," that is, open an empty file and put in all the necessary information to do the molecular mechanics of CH_4. What is the enthalpy of formation of CH_4? What are the C—H bond lengths and angles?

6 What is the enthalpy of formation of styrene by molecular mechanics?

7 What is the enthalpy of formation of phenylcyclopentane by molecular mechanics?

8 What enthalpy difference would lead to a 25–75% mix of *syn* and *skew* rotamers of 1-butene? Neglect any entropy change.

9 The double bond in Fig. 10-11a is at the bottom. If you did not know this, how could you tell, from the diagram alone?

REFERENCES

Allinger, N. L., 1976. *Adv. Phys. Org. Chem.*, **13**, 1.

Allinger, N. L., Dodziuk, H., Rogers, D. W., and Naik, S. N., 1982. *Tetrahedron*, **38**, 1593.

Atkins, P. W., 1986. *Physical Chemistry*, 3rd ed. W. H. Freeman, New York.

Burkert, U. and Allinger, N. L., 1982. *Molecular Mechanics*, ACS Monograph 177. American Chemical Society, Washington, DC.

Chong, D. P., 1985. QCMPE 004, Quantum Chemistry (Microcomputer) Program Exchange, Department of Chemistry, Indiana University, Bloomington, IN.

Clark, T., 1985. *A Handbook of Computational Chemistry*. Wiley, New York.

Cox, J. D. and Pilcher, G., 1970. *Thermochemistry of Organic and Organometallic Compounds*. Academic, London.

Fowles, G. R., 1962. *Analytical Mechanics*. Holt, Reinhart and Winston, New York.

Henkel, G. J. and Clarke, F. H., 1986. *Molecular Graphics on the IBM* ® *Microcomputer* 2.0 Diskette and Instruction Manual. Academic, San Diego, CA. Serena Software, P. O. Box 3076, Bloomington, IN. 47402-3076 also provides molecular graphics, MM2, and MMX molecular mechanics for pi systems, for IBM and compatible computers.

Morrison, R. T. and Boyd, R. N., 1973. *Organic Chemistry*. Allyn and Bacon, Boston, MA.

Pedley, J. B., Naylor, R. D., and Kirby, S. P., 1986. *Thermochemical Data of Organic Compounds*. Chapman and Hall, London.

Rogers, D. W., 1988. *Amer. Lab.*, **20** (10), 122.

Scott, L. T., Cooney, M. J., Rogers, D. W., and Dejroongruang, K., 1988. *J. Amer. Chem. Soc.*, **110**, 7244.

Saunders, M., 1987, 1990. *J. Amer. Chem. Soc.*, **109**, 3150; **112**, 1512.

van Hemeluijk, D., et al., 1980. *J. Amer. Chem. Soc.*, **102**, 2189.

11 | *Molecular Graphics*

Most scientists think pictorially. This is evident from the hundreds of graphs and diagrams that crowd our textbooks and the thousands of slides shown at our national meetings.

Pictorial presentation of information and its use in persuasion (as in propaganda and its close cousin, advertising) has enormous application outside of science and has been developed to a high degree. It is possible, for example, to program a computer to present pictures of three objects of different metals so that the viewer will immediately recognize that one is made of copper, one is of bronze, and one is pewter. These are not computer stored pictures of copper, bronze, and pewter objects: They are programs written to depict objects made of the three metals.

This chapter is an introduction to the complex and rich field of computer graphics.

TAB GRAPHICS

The simplest computer graphics technique, and the one that gives the crudest results, uses the TAB statement in BASIC. The TAB statement in BASIC is intended for the same use that the tab key has on a typewriter. If one wants to make a table with a column 20 spaces from the left margin, one sets the tab to 20 before printing the numerical result. Suppose the first value is 10. The print statement

$$X = 10$$
$$PRINT\ TAB(20)X$$

would do. The number in parentheses may be replaced by a variable so that

$$A = 20$$
$$X = 10$$
$$PRINT\ TAB(A)X$$

would achieve the same thing. Putting X in quotes causes it to be printed as an alphanumeric character. Thus PRINT TAB(A)"X" places the symbol

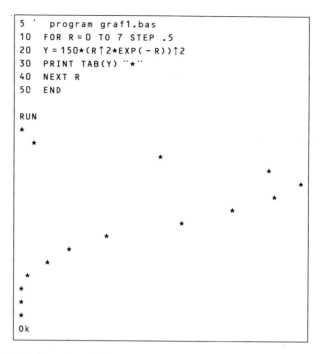

```
5 '   program graf1.bas
10  FOR R = 0 TO 7 STEP .5
20  Y = 150*(R↑2*EXP(-R))↑2
30  PRINT TAB(Y) "★"
40  NEXT R
50  END

RUN
★
   ★

              ★

                    ★
                      ★
                    ★
                 ★
            ★
         ★
      ★
    ★
  ★
 ★
★
★
★
Ok
```

FIGURE 11-1 TAB plot of the electron density in the $1s$ orbital of hydrogen as a function of radial distance from the nucleus.

X A spaces from the left margin. The same can be done with any alphanumeric character.

If A is set equal to some function, the function is evaluated and its value is stored in memory location A. The TAB causes an X to be printed at some distance from the left margin that is determined by the value of the function. This is what we do when we plot a point on a graph. If the TAB function is used repeatedly, as in a loop, several points can be plotted for different values of the independent variable (A is dependent) and one has a simple graph of the function. This is what has been done for Computer Graphs 11-1 and 11-2 (see Fig. 11-1).

Computer Graph 11-1

The lowest energy atomic wave function for hydrogen can be written

$$\psi_{1s} = e^{-r} \tag{11-1}$$

in rectangular coordinates or

$$\psi_{1s} = r^2 e^{-r} \tag{11-2}$$

in spherical polar coordinates. The variable r is the radial distance from the nucleus to any point in space. For simplicity in presentation, the functions are not normalized and the units or r are arbitrary. The square of the orbital ψ is the probability function for finding the electron at radius r. Program GRAF1 uses TAB graphics to plot $\psi^2 = f(r)$ in spherical polar coordinates for Eq. (11-2). Other orbitals or, indeed, any function can be substituted in statement 20 to compare the general curve shapes. The plot appears rotated 90° from the usual presentation but the method has the advantage of simplicity and is a quick way to look at functional behavior. The 150 that multiplies the function in statement 20 is a scaling factor that makes the plot "look right" on the page. It has no physical significance.

Computer Graph 11-2

Program GRAF2 (disk) works on the $2s$ orbital of hydrogen in spherical polar coordinates

$$\psi = r^2(2 - r)e^{-r} \tag{11-3}$$

in the same way as GRAF1 worked on the hydrogen $1s$ orbital. Note that these programs plot the squares of the wave functions which are nonnegative.

Pixel Graphics

One of the main drawbacks of TAB graphics—lack of resolution—can be reduced by using pixel graphics. A dot-matrix printer forms letters and numbers as arrays of small dots called pixels. The symbols X and *, therefore, use a block of pixels, perhaps as many as 24. By controlling each pixel individually or in small groups, much finer resolution is achieved. Thus a crude graph formed from 20 alphanumeric symbols in TAB graphics might be formed from several hundred individual pixels, or pixel groups, each representing a value of the function under investigation. GRAF3 (disk) plots the $2s$ orbital (squared) of hydrogen in pixel graphics (see Fig. 11-2).

Solutions of the function are carried out in the same way as in GRAF1 and GRAF2 except that there are more of them and they are closer together. The pixel graphics statement in GRAF3 is PSET(10 * R,Y)

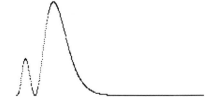

FIGURE 11-2 Pixel graphics of the electron density in the $2s$ orbital of hydrogen as a function of radial distance from the nucleus.

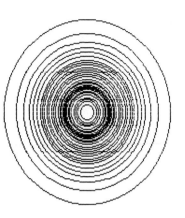

FIGURE 11-3 Contour diagram of the electron probability distribution of the $1s$ orbital of hydrogen. Line density is proportional to probability density. The horizontal images are artifacts of the printing process and would not appear in a perfect contour diagram.

which sets each pixel at a location 10R and Y on the CRT screen. The statement Y = 150-Y is used because the locations on the CRT screen are numbered from the top left-hand corner. The first variable (nominally X) goes from left to right in the normal way but the second variable of the number pair required by the PSET statement (nominally Y) is measured from top to bottom of the CRT screen. Thus the graph would come out upside down without the 150-Y statement. The 150-Y statement establishes a horizontal axis about 1/3 of the distance from the bottom of the screen and inverts the graph to its usual orientation. The nominal independent variable X is replaced by 10∗R in this program where 10 is a scaling factor to make the graph more legible.

GRAF4 produces the black-body radiation curve (Chapter 1) in pixel graphics.

Programs GRAF5 and GRAF6 (Figs. 11-3 and 11-4) take advantage of the simple spherical symmetry of s orbitals to plot contour diagrams of probability density for hydrogen. The area under the probability curve of ψ^2 as a function of R is integrated and each time the integral reaches a

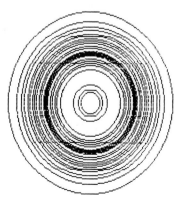

FIGURE 11-4 Contour diagram of the electron probability distribution of the $2s$ orbital of hydrogen. Line density is proportional to probability density. The horizontal lines are artifacts of the printing process and would not appear in a perfect contour diagram.

certain value, a circle is drawn. After the circle is drawn, the integral is set to 0 and the process is repeated. The integral builds up rapidly where the probability of finding an electron is high and builds up slowly with increments in R where the probability of finding an electron is small. Many circles are drawn where the electron is likely to be and few are drawn where it is not likely to be. The "shell structure" of hydrogen and hydrogen-like atoms is clear in the resulting contour diagram. This procedure can be carried out for atoms and molecules without spherical symmetry (Wahl, 1966; Streitwieser and Owens, 1973), but the programming is more difficult.

Statement CIRCLE (150,100),R is a special system subroutine that draws a circle of radius R with the center at point {150,100} measured from the top left corner of the CRT screen. The values chosen put the center of the circle about in the middle of the screen. In MS-DOS, GRAF5 and GRAF6 must be preceded by typing GRAPHICS at the system level and answering the questions on the kind of printer used. At the BASIC level, the preliminary command SCREEN 1 is necessary for pixel graphics. These diagrams were screen dumped to a good dot-matrix printer (Tandy DMP 430) using SHIFT PRINT.

PCMODEL

Many of the illustrations in this and succeeding chapters are skeleton or stick drawings made from screen dumps of two-dimensional projections of the MM2 structure using PCMODEL (Henkel and Clarke, 1986). These simple drawings were chosen to illustrate certain points of molecular geometry and do not represent the full range of techniques afforded by PCMODEL. Numerous other graphing and "docking" options are described in the PCMODEL documentation.

PCMODEL is an interactive program. After executing the PCMODEL.EXE file (type PCMODEL) and following directions indicated on the CRT, one arrives at a screen showing a blank quadrangle with a series of letters below it signifying command options. Command G activates a subroutine to *get* a new structure. After entering the answers to questions on how many atoms there are in the new structure, how many bonds, the x, y, and z coordinates of each atom, and the numbers of atoms that are connected by bonds, the new file is complete and is stored under a unique file name. For simple graphing, only two significant digits are necessary for the coordinates. If bond lengths and angles are to be calculated, one should maintain consistency in significant digits.

The new file (or any stored file) can be displayed to the CRT screen using the *add* command A. If the A command results in a blank screen, the diagram may be out of range of the viewing quadrangle on the CRT screen. It is usually necessary to *orient* the structure to the center of the screen using the O command. Numerous other commands exist for rotating, locating atoms, and adjusting the size of the diagram. After adjusting the CRT display as desired, the molecular diagram can be sent to the printer by the appropriate screen dump keystrokes. Screen dumps may require a graphics subroutine to be loaded (type GRAPHICS) after the system has been booted but before executing PCMODEL.

The coordinates from UNIT7 or OUTPUT2 of MM2 (Chong, 1985), used as input to PCMODEL, yield planar projections of any molecule from the coordinates generated in the MM procedure. The coordinate information necessary for PCMODEL plotting from MMX (Appendix) is found in FILENAME.OUT where FILENAME refers to the molecule under study.

Methane

Using the ethene file as a template, one can construct the input file for an MM calculation of the structure of methane. We shall use, as the starting geometry, the planar structural representation with carbon at the origin, hydrogen atoms 1.5 Å above and below on the y axis, and hydrogens 1.5 Å to either side on the x axis. By now, the rest of the input file should be familiar. This is a poor starting geometry; the bonds are too long and the planar structure is wrong.

The starting methane structure can be relaxed to its minimum energy using either MM2 or MMX (Appendix). The molecule relaxes away from the starting geometry with an energy reduction from 246 to 201 kcal mol^{-1}. One would not anticipate any significant strain in methane. This cannot be the true geometry because the energy is far too high for a small, unstrained molecule.

Inspection of the file OUTPUT2 shows that the model geometry has remained planar and minimization has only shortened the bond lengths, leading to partial reduction of the energy through relaxation of the stretched C—H bonds. The enthalpy of formation is unreasonably high (191 kcal mol^{-1}): Presumably, this is not a stable conformer but a structure at a saddle point on the potential energy surface.

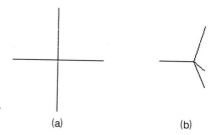

FIGURE 11-5 (a) Planar input for methane and (b) tetrahedral geometry resulting from minimization.

(a)   (b)

To "push" the model off its saddle point, one can modify the input file by giving a slight z component (0.1 Å) to one of the hydrogen atom positions. Rather remarkably, the initial steric energy is reduced to 225 kcal mol^{-1} by this small change. MMX or MM2 calculation yields -10 kcal mol^{-1}, a poor approximation to the experimental value of -17 kcal mol^{-1} but indication that relaxation has taken place and that the molecule is no longer at a saddle point. Strain energy is small; the discrepancy with the experimental value probably lies in the parametrization of the C—H bond enthalpy in methane.

The initial geometry, loaded into a PCMODEL data file *via* the G command yields Fig. 11-5(a). If corrections need to be made, use E for edit. After the coordinates are loaded and the file is recorded on disk, it can be oriented using the O command. Three atoms must be specified in response to program prompts to establish the atom, axis, and plane of orientation. When the image is center screen, it can be rotated (R command), translated to another part of the screen (T command), and increased in size (S command). See PCMODEL documentation for other options. If visual inspection of the image shows that something is wrong with it, it can be reedited using the E command. Visual inspection may also show something wrong with the original MM calculation and is an aid in distinguishing among close-lying conformers (Computer Project 11-2). Molecular mechanics and graphical modeling are best used in conjunction with one another.

One can copy the final coordinates, after MM relaxation, into a second file of PCMODEL and put both files on the same screen by adding the second file to the first. This is done by repeating the Add procedure while the first file is on the screen. The screen dump of the combined diagram shows both starting and final geometries. The result is Fig. 11-5. The expected tetrahedral structure of methane [Fig. 5-11(b)] supports our conclusion that a normal relaxation has taken place.

GRAPHING PROGRAMS

There are many commercial scientific graphing programs available. Each has as its object production of graphs and illustrations, as distinct from the specialized molecular modeling programs we have already seen. The most general graphics packages produce *x-y* plots such as *P-V* curves for gases, C_p-T curves for pure compounds, and so on. Specific-purpose graphics or drafting programs exist for chemistry, which produce molecular formulas like those seen in organic chemistry books. We shall distinguish among *plots*, which are single curves, *graphs*, which consist of one or more plots within a coordinate system, and *compositions*, which are two or more graphs grouped on a single page.

We shall introduce general-purpose graphing programs using SIGMAPLOT by Jandel Scientific as an example. SIGMAPLOT (Appendix) is a complex program with many options for the operator to chose from in constructing a graph.

Use of SIGMAPLOT is greatly simplified by its menu-driven structure. The SIGMAPLOT menu system enables one to determine where one is within the structure of the program, what options one has, and, if the desired option is not directly accessible, how to get to the portion of the program where it is. For example, at the edit menu, one may enter data for a plot, but to store the resulting plot, one must go to the disk menu. The reason there are several menus is that there are so many options. What appears to be a drawback to the beginner—nine separate menus—soon turns into its main strength. There are so many options that whatever kind of plot you want can probably be made.

The flow chart of menus gives an idea of the overall structure of the program:

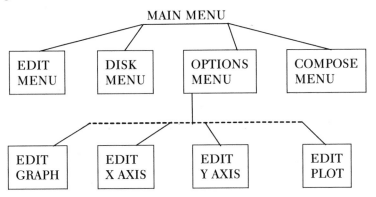

At any point in the flow chart, hitting the ESC key moves up in the chart. Thus, one can always move to the main menu and choose a path that arrives at the specific menu that has the option one wishes to execute. Jandel supplies very complete documentation so that you can locate any option within the flow chart.

Illustration: C_p vs. T for Lead

We have already integrated under the curve of C_p/T vs. T to determine the absolute entropy of lead in Computer Project 2-4. We can use the data in program ENTROPY to plot the curve of C_p vs. T using SIGMAPLOT. At the edit menu (keystroke F3 from the main menu), enter the first few points for T: $0, 5, 10, 15, 20$ in the first column of the quadrangle offered to you. In the second column, enter the heat capacities (multiply T into the first five entries for C_p/T in the DATA statement of program ENTROPY). The values are $0, 0.31, 2.8, 7.0, 10.8$. The data cursor

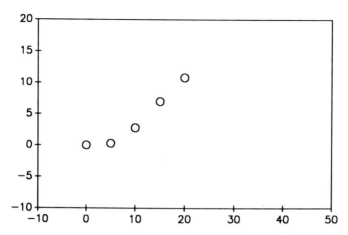

FIGURE 11-6 The first five points in a heat capacity vs. *T* plot for lead. (Extended graphics hardware.)

can be moved using the arrow keys. Hit ESC at the end of data entry. At the edit menu, title (F5) column 1 T and column 2 C. Designate the horizontal axis (col 1) and vertical axis (col 2) using the "pick plot" option, F4 at the main menu. Hit ESC at the completion of each operation. F7 from the main menu gives you a CRT presentation of the plot. If you have executed GRAPHICS from DOS, you can screen dump the plot to a printer (Fig. 11-6). If you wish to save the plot go to the disk menu and execute the "save graph" option.

Figure 11-6 shows the first five points of the data set. Upon entering all remaining data points at the edit menu, Fig. 11-7 is produced. The *x* range, − 100 and 500 in Fig. 11-7, is computed from the input data by the program. The *y* range − 10 to 50, is computed also. Both are rather larger

FIGURE 11-7 Heat capacity vs. *T* for lead. The *x* and *y* axial ranges are too large and will be changed in the next plot (Fig. 11-8). (Standard graphics hardware.)

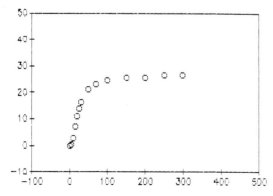

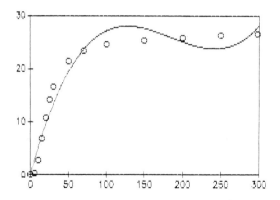

FIGURE 11-8 Heat capacity of lead as a function of T. The third-order regression curve gives a poor fit.

than we would like them, so we can go to the options menu and change the x range to 0 to 300 and the y range to 0 to 30 (option 2B). The tick label size is the interval you want between designators on the axis. This gives Fig. 11-8, but without the curve drawn through the points.

To draw the curve in Fig. 11-8, choose edit plot (F5) from the options menu. Option 7B permits us to select a regression order (Chapter 5 of this book). A third-order regression yields Fig. 11-8 with the curve drawn in. The fit is poor. Exercising option 7B once again, but this time selecting a fifth-order regression, yields Fig. 11-9, in which the fit is much better but still has regions of poor fit at the steep portion of the curve and what might be mistaken for a printer artifact near 280 K. These are shortcomings of the curve fit. We can see this by running a sixth-order regression whereupon the error at 280 K becomes much worse. Experi-

FIGURE 11-9 Heat capacity of lead as a function of T. The fifth-order regression curve gives a good fit.

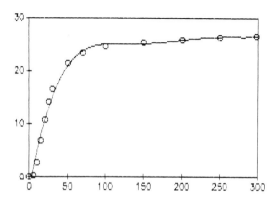

mental points can be fitted to any desired degree of accuracy by *some* function (Chapter 5), but merely increasing the order of a linear regression does not necessarily improve the fit.

Axes can be labelled by exercising further options within SIGMAPLOT. Graphs can be made smaller, the *x-y* ratio can be changed, multiple plots can be put on the same graph, or multiple graphs can be put on the same page using the *compose* option. For these options, see the program documentation. Computer hardware greatly influences the quality of the graph as can be seen by comparing Figs. 11-6 and 11-7.

COMPUTER PROJECT 11-1 | *Structure of Cyclopentene*

Determine the enthalpy of formation and structure of cyclopentene (Computer Project 10-4).

Cyclopentene is not planar as benzene is. The stable conformation is called an *envelope* form with one carbon atom out of the plane established by the sp^2 carbons. What are the internal angles of the cyclopentene ring and what is the angle between the plane of the sp^2 carbons and the out-of-plane sp^3 carbon atom?

Procedure Use PCMODEL to help in working this problem. The command V gives the angle among atoms designated. There are two options, S and T. The S option is used to determine the *simple* angle formed by three atoms. Use this option to determine the internal angles in the cyclopentene ring. The T option is used to determine the *torsional* angle. We have used the term *dihedral angle* to designate this angle. The terms can be considered synonymous for our purposes. Use the T option to determine the angle between the out of plane carbon and the planar carbons. You may wish to use the N option to determine which atoms on the CRT screen coincide with which numbers so as to interpret your dihedral angle. You should already have a pretty good idea of which atoms are which because you used the O option to orient the structure. How long is the double bond in cyclopentene? Use the D option to determine this. Compare your results for ΔH_f, internal angles, double bond length, and dihedral angle with the literature (Allinger et al., 1982) and with the output of Computer Project 10-4.

COMPUTER PROJECT 11-2 | *The Bicyclooctane System*

Bicyclo[3.3.0]octane has the structure one gets by fusing two pentagons so that they share a common side (Fig. 11-10). The fusion can be either cis or trans. The trans form is highly strained "mainly because of the very wide CCC inter ring angle of 124° by MM2" (Burkert and Allinger, 1982). Use MM calculations to determine the *cis-trans* isomerization enthalpy. Both

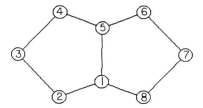

FIGURE 11-10 Numbering diagram for the bicyclo[3.3.0]octane system.

isomers are isolable, and combustion calorimetry has been carried out (Cheng et al., 1970), which yields an experimental result of 6.4 ± 0.8 kcal mol^{-1} for the isomerization enthalpy.

The *cis* form, being two "envelope" conformations, can have the "flaps" of the envelopes both down, both up, or one up and one down, as in Fig. 11-11. Which is the most stable? There is little difference in enthalpy among the conformers; the MM2 enthalpies of formation of conformers in Fig. 11-11 are -21.5, -22.5, and -22.8 kcal mol^{-1}.

Procedure Carry out the MM calculations for all three conformers. Load the coordinates of the MM calculation into PCMODEL and examine the geometries using the M command (for circular *motion* of the molecule). The structure must be centered on the CRT screen or the M command will send it off screen. Use this visualization to associate each enthalpy of

FIGURE 11-11 Three conformers of cis-bicyclo[3.3.0]octane. **(c)** and **(d)** are different views of the same conformer.

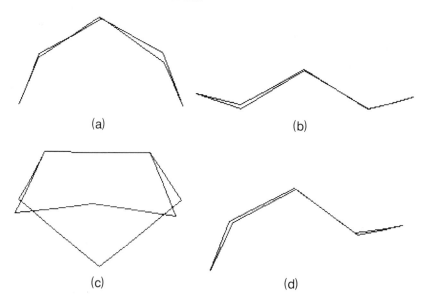

(a)

(b)

(c)

(d)

formation above with the appropriate conformer. Alternatively, assignment of ΔH_f values to conformers can be made by inspection of the dihedral angles.

One is tempted to regard part of the diagram as coming forward and part as receding into the background, but these diagrams are orthogonal projections of a three-dimensional figure (not perspective drawings), which have no foreground or background. PCMODEL has a feature that helps in this respect. Parts of the three-dimensional model that are away from the viewer are green and parts near the viewer are red.

COMPUTER PROJECT 11-3 | *Bicyclooctenes*

A number of interesting compounds can be drawn by inserting one or more double bonds in the bipentagonal parent structure. Both compounds with the double bond at the bridgehead, one bonded across the bridge, $\Delta^{1,5}$ bicyclo[3.3.0]octene, and the Bredt compound $\Delta^{1,2}$ bicyclo[3.3.0]octene are known and isolable (Greenberg and Liebman, 1978).

All of these bicycloalkenes are more rigid than the corresponding bicycloalkane. The monoalkene has a planar unsaturated ring and an envelope saturated ring with the flap up or down. The up position is the stable form (Fig. 11-12).

Double bonds in the 2 and 6 positions cause the pentagonal rings to warp in opposite directions away from their common bond of fusion and to assume the shape of an airplane propeller [Fig. 11-13(a)]. Figures 13(b) and (c) show the propeller of the *cis*-2,6-diene after a quarter turn about its vertical axis and after one-eighth turn, respectively. The pitch of the propeller can be determined from the dihedral angle of either pair of bonds *cis* to the bond of fusion, that is, the dihedral angle of carbons 2 1 5 4 or carbons 6 5 1 8. The dihedral angle 2 1 5 4, for example, gives the angle formed by bonds 1–2 and 5–4 when seen by an observer looking down the 1–5 bond axis. In the case of the *cis*-2,7-diene [Fig. 11-13(d)], this angle is small, but in the case of the *cis*-2,6-diene, the angle, and consequently the "pitch" of the propeller, is much larger.

Procedure Carry out a minimization of the energy of the 2,6- and 2,7-dienes using MM2 or MMX. Load the coordinates into PCMODEL. Use the V command to examine simple angles. Use the T command to examine the torsional (dihedral) angles of the diene structures. Report the pitch of the "propeller" form of both dienes in degrees.

FIGURE 11-12 *Cis*-bicyclo[3.3.0]octa-2-ene.

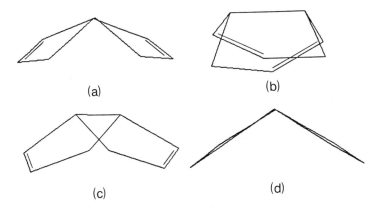

FIGURE 11-13 (a)–(c) Cis-bicyclo[3.3.0]octa-2,6-diene and (d) cis-bicyclo[3.3.0]octa-2,7-diene.

Presentation Graphics For the purpose of examining the architecture of molecules, the simple line drawings used so far are ideal, because they show the essentials of structure in an uncluttered way. For many cases, only the carbon skeleton is used, for exactly this reason. When preparing drawings or slides for presentation to a general audience, however, these structural models are too sparse. Persons unfamiliar with structural chemistry usually prefer models closer to the "ball-and-stick" models from general chemistry.

To produce ball-and-stick models from the structures we have generated, one must first add the hydrogen coordinates after the carbon atoms (PCMODEL files can be edited and added to) and execute keystroke F3. The resulting drawing is shown as Fig. 11-14, cyclopentane. The plane of the ring is not the same as the plane of the paper; hence, there is some foreshortening as well as some ring nonplanarity (fig. 11-15). PCMODEL permits rotation to any desired orientation before screen dumping. Use

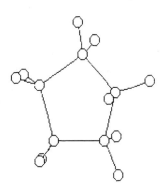

FIGURE 11-14 Ball-and-stick model of cyclopentane.

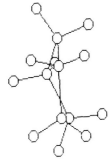

FIGURE 11-15 Ball-and-stick model of cyclopentane showing the envelope structure.

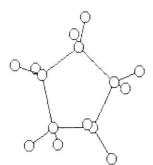

FIGURE 11-16 Figure 11-14 after use of the rotation option to obtain a slightly better presentation angle.

ALT E before the screen dump to expand the screen for a better presentation drawing (Fig. 11-16).

COMPUTER PROJECT 11-4 | *Entropy of Silver*

Plot C_p (equal to C_v for a metal) vs. T for the lustrous metal in Computer Project 2-4. The lustrous metal is silver. Frequently, C_p is plotted against ln T or log T. These plots have the advantage that the integral of C_p vs. ln T (or log T) yields the absolute entropy of a metal undergoing no phase change between 0 K and the desired temperature, usually 298 K. Instead of integrating numerically, as in Chapter 2, we shall obtain a high degree regression of the function and integrate analytically from 0 to 298 K.

Procedure Using SIGMAPLOT, plot the curve of C_p vs. T for silver as was done for lead earlier in this chapter. The data set is given in Computer Project 2-4. Store the resulting graph under the name SILVER. Calculate ln T for each temperature and load it into column 1 of the data set at the edit menu. Delete the old values of T. Plot C_p vs. ln T for silver.

Exercising option 7B, obtain the second-order regression for the plot. The regression equation is of the form

$$y = a_0 + a_1 x + a_2 x^2 + \cdots + a_n x^n$$

where n, the highest power of x, is the order of the regression, which is 2 in this case. Now exercise the STATS option (F6). This yields a listing of the coefficients a_0, a_1, and a_2 for the regression. Since $C_p = y$ and $\ln T = x$, the integral

$$\int_0^y y\, dx = \int_0^{\ln 298} C_p\, d(\ln T)$$

yields the absolute entropy as discussed in Chapter 2. The integral yields a sum of contributions to the entropy, one for each term in the regression polynomial. List these terms separately and obtain their sum. Comment on their relative size. Compare the entropy determined here with the entropy determined in Computer Project 2-4, which was 41.39 J K^{-1} mol^{-1}. The experimental value is 42.55 J k^{-1} mol^{-1}.

Repeat the integration procedure for a fourth-degree regression. Comment on the results.

Try higher orders. Notice that some regressions lead to a large deviation below $\ln T = 1$. These can be avoided by integrating only over the interval for which the integral is positive: $\ln T = 2.2$ to 5.7 in the second-order case. The entropy contribution at temperatures near 0 K is very small and is usually calculated by the Debye extrapolation (Atkins, 1986). At the level of approximation of a second- or fourth-order polynomial fit, the Debye extrapolation is hardly justified, but it is important in more accurate work.

PROBLEMS | Chapter 11

1 Generate Fig. 11-1(b) using program GRAF2.

2 Screen dump the results of GRAF2 and determine the ratio of distances from the origin to each of the two peaks in the resulting figure. One interpretation of the Bohr radii of hydrogen is that they are the radii of maximum electron probability density. Review Bohr theory in an elementary general or physical chemistry book and decide what the ratio of the first two Bohr radii should be. Does it agree with the ratio obtained from GRAF2?

3 Expand the horizontal axis of GRAF2 for more accurate results. Repeat Problem 1 using the expanded axis.

4 Command D in PCMODEL presents the distance between any two atoms in a structure to either screen or printer. Use this option to determine the distance between hydrogens in methane.

5 Command V in PCMODEL presents the simple angle between any three atoms in a molecule. Use this option to determine the (well-known tetrahedral) bond angle between C—H bonds in methane. Use the option to determine the angle between any C—H bond and the line joining that hydrogen atom with any of its neighbors, i.e., between the bond and one of the edges of the tetrahedron.

6 Using PCMODEL, with a structure on the CRT screen, hit F3 to generate a ball-and-stick model.

7 Using PCMODEL, with a structure on the screen, hit ALT S to generate a space-filling model. This is best done with a simple structure.

8 (a) Is either of Eqs. (11-1) or (11-2) normalized in the quantum-mechanical sense? (b) Use the integral of the square of each wave function to determine a normalizing factor for that function (Chapter 2). (c) What is the probability of finding the $1s$ electron within 1.5 Bohr radii of the nucleus? (1 Bohr radius = 0.053 nm = 53 pm.) (d) Rotate (mentally) Figs. 11-3 and 11-4 out of the plane of the paper. This is the three-dimensional probability distribution for the $1s$ electron of hydrogen. How do the plots differ in cartesian and spherical coordinates? What is the graphical meaning of the term electron shell model of the atom?

9 Use the results of Computer Project 10-2 to produce a molecular diagram of the *cis* and *trans* 2-butenes. What is the distance between carbons 1 and 4 in the *trans* isomer? What is the distance between carbons 1 and 4 in the *cis* isomer?

10 How far apart are the methylene hydrogens in cyclopentene? Use the results from Computer Project 11-1.

11 Obtain the third-order polynomial regression of C_p vs. ln T for lead (Computer Project 2-4). Integrate and obtain the absolute entropy. Compare your result using analytical integration with the result from Simpson's rule of integration and the experimental value of 64.81 J K^{-1} mol^{-1} (Atkins, 1986).

REFERENCES

Allinger, N. L., 1976. *Adv. Phys. Org. Chem.*, **13**, 1.

Allinger, N. L., Dodziuk, H., Rogers, D. W., Naik, S., 1982. *Tetrahedron*, **38**, 1593.

Atkins, P. W., 1986. *Physical Chemistry*, 3rd ed. W. H. Freeman, New York.

Burkert, U. and Allinger, N. L., 1982. *Molecular Mechanics*, ACS Monograph 177. American Chemical Society, Washington, DC.

Cheng et al., 1970. *J. Amer. Chem. Soc.*, **92**, 3109.

Operating Instructions for MM2 and MMP2 Programs. 1977 *Force Field*. QCPE Manual for Program 395; QCMPE Manual for Program 004.

Clark, T., 1985. *A Handbook of Computational Chemistry*. Wiley, New York.

Gilbert, K. E. and Gajewski, J. J., 1988. Serena Software.

Greenberg, A. and Liebman, J. F., 1978. *Strained Organic Molecules*. Academic, New York.

Henkel, G. J. and Clarke, F. H., 1986. *Molecular Graphics on the IBM®️ Microcomputer 2.0*. Diskette and Instruction Manual. Academic, San Diego, CA.

Osawa, E., 1979. *J. Amer. Chem. Soc.*, **101**, 5523.

Rogers, D. W., 1987. *Am. Lab.*, **19** (4), 28.

Streitwieser, A. and Owens, P. H., 1973. *Orbital and Electron Density Diagrams*. Macmillan, New York.

Wahl, A. C., 1966. *Science*, **157**, 961.

12 | Molecular Quantum Mechanics: The ab initio Method

We have introduced certain ideas of molecular quantum mechanics as needed in Chapters 7 through 9. To go ahead in a more systematic way, we should recapitulate.

The Schroedinger equation

$$H\psi = E\psi \tag{12-1}$$

applies to all mechanical systems. The hamiltonian operator H differs for different mechanical systems. By analogy to the classical equation

$$T + V = E \tag{12-2}$$

where T is the kinetic energy and V is the potential energy, the hamiltonian operator H has a kinetic energy part and a potential energy part. For the hydrogen atom,

$$H = -\left[\tfrac{1}{2}\nabla^2 + V\right] \tag{12-3}$$

in atomic units. The part with ∇^2 is the kinetic energy part and V is the potential energy part.

For a system of many nuclei and many electrons,

$$H = \sum -\tfrac{1}{2}\nabla^2(i) - V(i) + \frac{1}{2}\sum \frac{1}{r_{i,j}} + \frac{1}{R} \tag{12-4}$$

where the $1/r_{ij}$ term arises through repulsion between one electron and another and $1/R$ is a nuclear repulsion term. The $1/r_{ij}$ term cannot be evaluated exactly because, in order to know the potential energy of electron i upon j, we must know the position of j, which depends on i.

If we expand ψ as a product of antisymmetrized orbitals (Chapter 9), say one for each electron, set the $1/R$ term aside for the time being, and drop the offending r_{ij} term, we have a set of n simultaneous equations in the n orbitals:

$$H(i) = \sum -\tfrac{1}{2}\nabla^2(i) + V(i) \tag{12-5}$$

The $H(i)$ are one-electron hamiltonians. This set of hamiltonians, substituted into the Schroedinger equation under the LCAO approximation leads to the secular equations

$$\sum (H_{rs} - E_i S_{rs}) a_{is} = 0 \tag{12-6}$$

There are many ways of compensating for the omission of the r_{ij} terms. One can simply ignore them, as in Huckel theory (Chapter 7), and take the molecular wave function ψ as the product of one-electron wave functions

$$\psi = \psi(1)\psi(2) \cdots \psi(n) \tag{12-7}$$

In this theory, the secular matrix elements are not evaluated; α is set arbitrarily to zero (as a reference point) and energies are kept in units of β. The exchange integral β can be determined empirically (e.g., Computer Project 7-2), but the results are not very satisfactory.

In simple SCF theory, along with empirical substitution for matrix elements H_{rs}, an iterative method is used to correct the integrals of successive calculations according to the results of the previous calculation. This is repeated until the results are self-consistent. In this way, some correction for r_{ij} interactions is built into the procedure, because the iterative correction process builds up an average field V for the ith electron that includes all electrons that are not i. The self-consistent field method will be used throughout this chapter.

In atomic units, the hamiltonian operator for the hydrogen atom is

$$H = \left[-\frac{1}{2}\nabla^2 - \frac{1}{r} \right] \tag{12-8}$$

whence the Schroedinger equation

$$H\psi = E\psi \tag{12-9}$$

becomes

$$\left[-\frac{1}{2}\nabla^2 - \frac{1}{r} \right]\psi = E\psi \tag{12-10}$$

with the hydrogenic atomic orbitals as exact solutions.

The Schroedinger equation for the hydrogen molecule ion H_2^+ is

$$\left[-\frac{1}{2}\nabla^2 - \frac{1}{r_A} - \frac{1}{r_B} \right]\psi = E_{el}\psi \tag{12-11}$$

with the homonuclear molecular orbitals as exact solutions. The Born–Oppenheimer approximation permits us to calculate the electronic energy separately from the total energy which includes the nuclear repulsion term $1/R$ (always positive)

$$E_{total} = E_{el} + \frac{1}{R} \tag{12-12}$$

where R is the internuclear distance. When E_{total} is calculated for many values of R, $E_{\text{total}} = f(R)$ has a minimum at the equilibrium bond length of the H_2^+ ion.

Because it has only one electron (no $1/r_{ij}$ term), H_2^+ can be solved exactly. The method cannot be extended to many-electron systems, however. Hence, we shall be more interested in approximate methods, which can be extended.

Although we shall usually treat the wave function as though it were a function of three spatial dimensions, it is really a function of four, the fourth being a relativistic or spin dimension leading to a quantum number $+\frac{1}{2}$ or $-\frac{1}{2}$. We shall, when necessary, separate the wave function into a space part of three dimensions and a spin part of one.

H_2^+ BY THE LCAO METHOD

Under the LCAO approximation

$$\psi = a_1 e^{-r_A} + a_2 e^{-r_B}$$

because $a_i e^{-r}$ are normalized hydrogen wave functions. As before,

$$E = \int \psi H \psi \, d\tau \Big/ \int \psi^2 \, d\tau \qquad (12\text{-}13)$$

These integrals are often abbreviated using a bracket notation $\langle \ \rangle$ for simplicity:

$$E = \frac{\langle \psi | H | \psi \rangle}{\langle \psi | \psi \rangle} \qquad (12\text{-}14)$$

For normalized atomic orbitals, the denominator is an integral that has been set equal to 1 in the normalization process. By definition,

$$\langle \psi | \psi \rangle = 1$$

hence

$$E = \langle \psi | H | \psi \rangle$$

This leads to

$$E = \frac{1}{2(1 + S)} \left\langle (1s_A + 1s_B) \left(-\frac{1}{2} \nabla^2 - \frac{1}{r_A} - \frac{1}{r_B} \right) (1s_A + 1s_B) \right\rangle \qquad (12\text{-}15)$$

where S is the overlap integral and $1s_A$ and $1s_B$ are the hydrogenic atomic orbitals used as basis functions in the LCAO method, and $1/(2(1 + S))$ is the normalization coefficient, which has been factored out. The molecule–ion is symmetrical; hence,

$$\left\langle 1s_A\left(-\tfrac{1}{2}\nabla^2\right)1s_A\right\rangle = \left\langle 1s_B\left(-\tfrac{1}{2}\nabla^2\right)1s_B\right\rangle \tag{12-16}$$

and

$$\left\langle 1s_A\left(\frac{1}{r_A}\right)1s_A\right\rangle = \left\langle 1s_B\left(\frac{1}{r_B}\right)1s_B\right\rangle \tag{12-17}$$

If we expand Eq. (12-15) as a polynomial and simplify using Eqs. (12-16) and (12-17), we get

$$E = \frac{1}{1+S}\left(\left\langle 1s_A\left(-\frac{1}{2}\nabla^2\right)1s_A\right\rangle + \left\langle 1s_A\left(-\frac{1}{2}\nabla^2\right)1s_B\right\rangle\right.$$
$$\left. -\left\langle 1s_A\left(\frac{1}{r_A}\right)1s_A\right\rangle - \left\langle 1s_A\left(\frac{1}{r_B}\right)1s_A\right\rangle - 2\left\langle 1s_A\left(\frac{1}{r_B}\right)1s_B\right\rangle\right) \tag{12-18}$$

Evaluation of the integrals in Eq. (12-18) and inclusion of the internuclear distance R (Mc Quarrie, 1983) yields

$$E = \frac{J+K}{1+S} \tag{12-19}$$

where

$$J = e^{-2R}\left(1 + \frac{1}{R}\right)$$

$$K = \frac{S}{R} - e^{-R}(1+R)$$

$$S = \left(1 + R + \frac{R^2}{3}\right)e^{-R}$$

EXERCISE 12-1

The equilibrium internuclear distance in H_2^+ is 2.49 a.u. (atomic units) in this first approximation. Calculate the dissociation energy.

Solution 12-1

$$S = 0.461, \qquad J = 0.00963, \qquad K = -0.1065,$$
$$E = -0.066 \text{ a.u.}$$

The energy is negative, which indicates bonding. The dissociation energy necessary to separate the ion into its parts is positive by convention.

EXERCISE 12-2

What percentage of the experimental dissociation energy (Computer Project 2-6) is obtained in this first approximation?

Solution 12-2 Multiplication of the calculated dissociation energy by the conversion factor from atomic units to kilojoules per mole (2625) yields 173 kJ mol^{-1}. This is 67% of the experimental value. Inclusion of variational parameters in the exponents of the $1s$ orbitals improves the calculation so that it yields 86% of the experimental value. Taking more basis functions permits one to refine the calculation so that it converges on the experimental value to within any desired uncertainty limits.

The Molecular Energy Manifold

Along with the negative solution Eq. (12-19), there is one that leads to a positive energy (energy increase)

$$E = \frac{J - K}{1 - S}$$
$$= 0.215 \text{ a.u.} \qquad (12\text{-}20)$$

at an internuclear distance of 2.49 a.u. in this approximation. The lower molecular orbital is called $1\sigma_g$ because it is unchanged by inversion through its center of symmetry (gerade: even) and the other is $1\sigma_u$ because it changes sign when inverted through its center of symmetry (ungerade: odd).

If the $2s$ orbitals are taken into the LCAO, four molecular orbitals result. If a larger basis set is taken, more MOs result, leading to the familiar σ-π energy level manifold discussed in physical chemistry or general chemistry texts. The basis set may, but need not, consist of atomic orbitals (AOs).

The Hydrogen Molecule

The hamiltonian operator for H_2 is

$$H = -\frac{1}{2}\nabla_1^2 - \frac{1}{2}\nabla_2^2 - \frac{1}{r_{1A}} - \frac{1}{r_{1B}} - \frac{1}{r_{2A}} - \frac{1}{r_{2B}} + \frac{1}{r_{12}} \qquad (12\text{-}21)$$

where the subscripts 1 and 2 refer to the electrons and A and B refer to the nuclei. This is just the hamiltonian for H_2^+ for electron 1, plus the hamiltonian for H_2^+ for electron 2, plus the interelectronic interaction term $1/r_{12}$:

$$H = H_{(1)} + H_{(2)} + \frac{1}{r_{12}} \qquad (12\text{-}22)$$

Each one-electron hamiltonian leads to a one-electron equation, as in Eqs. (12-5):

$$H_{(1)}\psi_1 = E_{(1)}\psi_1$$
$$H_{(2)}\psi_2 = E_{(2)}\psi_2$$

(12-23)

neither of which is the exact solution for H_2 because of omission of the $1/r_{12}$ term. Each equation leads to a one-electron energy E_i that is part of the electronic energy.

The Pauli Principle

One statement of the Pauli exclusion principle (Chapter 9) is that "no two electrons in a system can have all four quantum numbers the same." Translated into molecular terms, this means that either the spin part of the total wave function must be antisymmetric (electrons paired, wave function g: even) or the space part must be antisymmetric (electrons unpaired, wave function u: odd).

The first situation corresponds to the *ground state* for hydrogen,

$$\psi = 1\sigma_g(1)\overline{1\sigma_g(2)}$$

(12-24)

in the LCAO approximation, where

$$\sigma_g = \frac{1}{\sqrt{2}}(1s_A + 1s_B)$$

(12-25)

for the two electrons 1 and 2 α or β where the overbar indicates the opposed electron spin.

In solving for the ground-state energy (Richards and Cooper, 1983), one is led to two times the one-electron energy plus an added term,

$$J = \left\langle 1\sigma_g(1)\overline{1\sigma_g(2)} \left(\frac{1}{r_{12}} \right) 1\sigma_g(1)\overline{1\sigma_g(2)} \right\rangle$$

(12-26)

that accounts for the r_{12} interactions. This integral yields an energy contribution of J; hence, the ground-state energy of H_2 is

$$E = 2E + J$$

(12-27)

for the electronic energy or

$$E = 2E + J + \frac{1}{R}$$

(12-28)

for the total energy of the molecule.

Dewar (1971) has shown how the space and spin parts factor for the ground-state solution of H_2 and how integration over the spin part produces only 0 or 1 for orthonormal basis functions. The result of this simplification is that the K integrals, which we might expect by analogy to H_2^+, do not appear in the ground-state solution.

Note the separation of the energy into one-electron integrals E and two-electron integrals represented here by J. In a more complicated system than H_2 there will be more one-electron integrals, but there will be very many, perhaps millions, of two-electron integrals. That is why *ab initio* calculations were not widely used until the advent of powerful computers.

The H_2 Excited State

The second molecular orbital for the two electrons in hydrogen is

$$\psi = 1\sigma_g(1)1\sigma_u(2) \qquad (12\text{-}29)$$

where the lack of an overbar shows that the electrons are unpaired and

$$\sigma_u = \frac{1}{\sqrt{2}}(1s_A - 1s_B) \qquad (12\text{-}30)$$

This is the *excited* or triplet state of H_2.

When the spin part of the wave function is symmetrical, as in the excited state of hydrogen, the space part of the wave function must be antisymmetrical,

$$H = \frac{1}{2}\left\langle 1\sigma_g(1)1\sigma_u(2) - 1\sigma_g(2)1\sigma_u(1)\left(H_{(1)} + H_{(2)} + \frac{1}{r_{12}}\right)\right.$$

$$\left. \times 1\sigma_g(1)1\sigma_u(2) - 1\sigma_g(2)1\sigma_u(1)\right\rangle \qquad (12\text{-}31)$$

When this expression is expanded, some of the integrals that dropped out in the ground state because of spin orthogonality

$$\langle \alpha(1)\beta(1)\rangle = 0$$

no longer drop out because in the excited state, the spins are the same; that is, they are normal functions

$$\langle \alpha(1)\alpha(1)\rangle = 1$$

These are the K or *exchange* integrals

$$K = \left\langle 1\sigma_g(1)1\sigma_u(1)\left(\frac{1}{r_{12}}\right)1\sigma_g(2)1\sigma_u(2)\right\rangle \qquad (12\text{-}32)$$

The K integrals are two-electron integrals as are the J or *coulomb* integrals [Eq. (12-26)]. The electronic energy is the sum

$$E = 2E + J - K \tag{12-33}$$

for H_2.

The Hartree–Fock Equations

When larger basis sets are taken, as they inevitably must be for molecules larger than H_2, it is the task of the computer programmer to write programs for evaluation of all the one- and two-electron energies and to obtain the corresponding wave functions.

For closed shell systems (all electrons paired), one has for the general case of many electrons

$$E = \sum 2E_i + \sum (2J_{ij} - K_{ij}) \tag{12-34}$$

where i and j are running indices over the orbitals. Each one-electron energy is related to an approximate operator

$$\left[\mathbf{H} + \sum \mathbf{J}_j - \sum \mathbf{K}_j\right]\psi_i(1) = E_i\psi \tag{12-35}$$

The $\mathbf{J}_j$ and $\mathbf{K}_j$ are operators, each corresponding to the integrals J_{ij} and K_{ij} in the way that $\mathbf{H}$ corresponds to E in the exact Schroedinger equation; that is, when the $\mathbf{J}$ and $\mathbf{K}$ operators operate on the wave function, the result is an energy times the wave function. In the case of the $\mathbf{J}_j$ and $\mathbf{K}_j$ operators, the energies are the coulomb and exchange energies, respectively.

All of the operators, $\mathbf{H}$, $\mathbf{J}$, and $\mathbf{K}$ can be summed to give the *Fock operator*, $\mathbf{F}$:

$$\mathbf{F}\psi_i(1) = E_i\psi_i(1) \tag{12-36}$$

where the equations are to be solved by a self-consistent field method as discussed in Chapter 9. This indeed is the equation solved in Chapter 9 [Eq. (9-18)], without the empirical elements of the H matrix that we used there. Without assumed matrix elements, this is one form of the Hartree–Fock equation for closed systems. If this equation is solved, one is said to have arrived at the Hartree–Fock limit. The Hartree–Fock limit is not an exact solution, because the Hartree–Fock equation is not the Schroedinger equation, involving, as it does, assumptions about each electron moving in the *average* field of all the other electrons.

If one makes the LCAO approximations for the wave functions in the Hartree–Fock equations

$$\phi_i = \sum a_j\phi_j$$

one obtains the Roothaan equations

$$\mathbf{F}\sum a_{ij}\phi_j = E\sum a_{ij}\phi_j \tag{12-37}$$

which are simultaneous equations in the minimization parameters a_{ij}. The normal equations [see Eq. (8-1)]

$$(\mathbf{F} - E\mathbf{S})\mathbf{a} = 0 \qquad (12\text{-}38)$$

may be solved by iterative diagonalization as described in Chapters 7 and 9, but the summed integrals are no longer replaced by empirical parameters; they are solved individually.

Notation

Up to this point notation has been nonrigorous and simplified as appropriate to a general presentation. We have attempted not to interrupt the logical flow of the presentation with frequent comments on notation, which can be difficult in this field. Rigorous notation will be found in the more advanced treatments (e.g. Hehre et al., 1986) given at the end of this chapter. The Roothaan equations are written

$$\sum (F_{\mu\nu} - E_i S_{\mu\nu}) a_{\nu i} = 0 \qquad (12\text{-}39)$$

This is essentially Eq. (9-18).

The Greek epsilon (ϵ) is often used to designate the energy and c is often used as the minimization parameter that we have called a. By one convention, molecular orbitals and basis functions are differentiated by using Roman subscripts for the former and Greek subscripts for the latter; thus,

$$\psi_i = \sum A_{\mu i} \phi_\mu$$

is the expansion of molecular orbital ψ_i in the basis set ϕ_μ, with the expansion coefficients $a_{\mu i}$.

The normalization conditions are

$$\sum \sum a_{\mu i} S_{\mu\nu} a_{\nu i} = 1 \qquad (12\text{-}40)$$

The overlap matrix element $S_{\mu\nu}$ is

$$S_{\mu\nu} = \langle \phi_\mu(1) | \phi_\nu(1) \rangle \qquad (12\text{-}41)$$

The elements of the Fock matrix are

$$F_{\mu\nu} = H_{\mu\nu}^{\text{core}} + \sum \sum P_{\lambda\sigma} [\langle \mu\nu | \lambda\sigma \rangle - \tfrac{1}{2} \langle \mu\lambda | \nu\sigma \rangle] \qquad (12\text{-}42)$$

$H_{\mu\nu}^{\text{core}}$ is a matrix element representing one electron in the field of all nuclei

$$H_{\mu\nu}^{\text{core}} = \langle \phi_\mu(1) | H^{\text{core}} | \phi_\nu(1) \rangle$$

$$\mathbf{H}^{\text{core}}(1) = -\frac{1}{2} \nabla^2 - \sum \frac{Z_A}{r_{1A}}$$

The quantities $\langle \mu\nu|\lambda\sigma \rangle$ are two-electron repulsion integrals

$$\langle \mu\nu|\lambda\sigma \rangle = \int \phi_\mu(1)\phi_\nu(1)\left(\frac{1}{r_{12}}\right)\phi_\lambda(2)\phi_\sigma(2)\, d\tau \qquad (12\text{-}43)$$

The electronic energy is

$$E_{el} = \tfrac{1}{2}\sum\sum P_{\mu\nu}\left(F_{\mu\nu} + H_{\mu\nu}^{core}\right) \qquad (12\text{-}44)$$

where

$$P_{\lambda\sigma} = 2\sum c_{\lambda i} c_{\sigma i} \qquad (12\text{-}45)$$

is an element of the density matrix P. Elements of the density matrix are related to the charge density at an atom $P_{\lambda\lambda}$ or between atoms $P_{\lambda\sigma}$ (Hinchliffe, 1988). The total energy is

$$E = E_{el} + E_{nuc\ rep} \qquad (12\text{-}46)$$

where, as before, the nuclear repulsion energy is added to E_{el}. These are the orbital energies that are obtained by SCF iteration of any basis set. If the basis set is sufficiently large that no change occurs upon arbitrary expansion of the basis set, the Hartree–Fock limit has been reached. The Hartree–Fock limit is not an exact solution of the Schroedinger equation because of the approximations made in the basis set and derivation of the Roothaan equations.

Slater-Type Orbitals (STO)

In the LCAO approximation, we wish to take a linear combination of atomic orbitals to represent each molecular orbital. If these are the radial hydrogenic atomic orbitals,

$$s: \quad Ne^{-zr}$$
$$p: \quad N_1 re^{-zr}$$
$$d: \quad N_2 r^2 e^{-zr}$$

they are called Slater-type orbitals. If we take one STO for each molecular orbital, the basis set is called a *minimal basis set*. Better results can be obtained by increasing the basis set. One way of doing this is to take two basis functions for each molecular orbital, each with a different value of zeta. This set is called a *double zeta* (DZ) basis set. The value of zeta in the double zeta basis set is often obtained by minimization of the molecular energies or, if this is not practical, minimization of the atomic energy (Richards and Cooper, 1983).

Gaussian Basis Sets (GTO)

Slater-type orbitals, which have a radial part $r^{(n-1)}e^{-zr}$, are not easily handled by computer, but gaussian orbitals, which have a radial part

e^{-zr^2}, are. In particular, the product of two gaussians is a gaussian, whereas the same is not true of STOs. It is currently almost universal practice to use gaussians in *ab initio* basis sets, as we shall do in the exercises. The simplest way to do this is to approximate the STO that one would like to use with a linear combination of gaussians. The smallest number of gaussians that give a good approximation to an STO is three; hence, the simplest basis set in common use is the STO-3G basis set. This is the basis set that we shall use in the exercises.

The exponents in a gaussian basis set are fitted parameters (Chapter 5) such that the STO is best approximated. The linear combination is

$$\psi = d_1 e^{-a_1 r^2} + d_2 e^{-a_2 r^2} + d_3 e^{-a_3 r^2} \tag{12-47}$$

where the d_i are the contraction coefficients and a_i are the exponents (strictly, exponential coefficients). Each of the terms on the right of Eq. (12-47) is called a *primitive*.

Contracted Gaussian-Type Orbitals (CGTO)

It is advantageous to expand the basis set beyond three gaussians. When this is done, STOs can be more closely approximated or, more to the point in contemporary calculations, the molecular orbitals can be improved without really worrying about whether the basis set does or does not approximate a set of STOs. In so doing, one is departing from the LCAO procedure.

In expanding the gaussian basis set, computational problems again increase. Huzinaga (1979) showed that gaussians in a large basis set can be broken up into groups and optimized as a group, rather than as individual functions. Computational problems are again mitigated by this strategy, but almost all of the advantage of the larger basis set is retained. These grouped basis sets are also said to be *contracted*. Different contraction schemes lead to different designations of basis sets such as 6-21G and larger *split valence* basis sets (Hehre et al., 1986).

In summary, the minimum STO basis set is expressed as a linear combination of three or more gaussians, STO-3G or greater. Each gaussian has a fitted exponential parameter (or two in DZ). When larger basis sets are used, they are grouped to reduce computation time. These contracted combinations lead to another set of parameters, the contraction coefficients.

THE COMPUTATION

The computational task is to compute all the one- and two-electron energies in Eq. (12-46) and to sum them with the internuclear repulsion energy so as to obtain the total energy of the molecule. This can be done for one input geometry in what is called a *single point* calculation. Conversely, the calculation can be carried out many times so as to approach the lowest energy for the system in a manner similar to a molecular mechanical minimization. Because, in the process of seeking the energy minimum, programs calculate the slope of approach (the energy gradient) this second, better, and more time-consuming calculation is called a gradient calculation.

Computation of the energy also generates the molecular wave function and the basis set coefficients (eigenvectors) as it did in the simple Huckel and SCF calculations. With these, it is a simple matter to obtain electron densities, bond orders, and dipole moments as was done in the earlier procedures.

Ab Initio Calculation of the Bond Energy of H_2

A simple molecular orbital calculation of the bond energy of H_2 (Coulson, 1937; Dewar, 1971) yields 0.099 a.u. of energy. This is 2.70 eV or 260 kJ mol^{-1}. The bond energy as a function of internuclear distance R has a minimum at 1.61 a.u. = 85 pm = 0.85 Å. The experimental value of bond (dissociation) energy is 0.174 a.u. = 456 kJ mol^{-1}. The Hartree–Fock

FIGURE 12-1 CADPAC input file for single-point *ab initio* calculations on H_2. The internuclear separation is 1.40 a.u.

```
C > type b:source.dat
  TITLE
      MINIMUM BASIS FOR h2
  MAXIT 20
  PRINT 3
  CONVERGENCE 7
  ATOMS
  hydrogen      1.          0.              0.      0.
      1    S    3
      1    3.425251  0.154329
      2    0.623914  0.535328
      3    0.168855  0.444635
  END
  HYDROGEN      1.0     2.40      0.0   0.0
      1    S _  3
      1    3.425251  0.154329
      2    0.623914  0.535328
      3    0.168855  0.444635
  END
  END
  START
  FINISH
  C >
```

limit is 0.1336 a.u. = 351 kJ mol^{-1}. Both experimental and Hartree–Fock limit energy minima are at an internuclear distance of 1.40 a.u. = 73.9 pm.

We shall carry out a single-point STO-3G calculation of the dissociation energy of H_2. This means that the calculation will be carried out at only one internuclear distance, 1.40 a.u., and that a minimum basis set of three gaussians will be used to approximate the Slater-type orbitals of the two hydrogen atoms in H_2.

The input file is given in Fig. 12-1. The *ab initio* program is MICRO CADPAC (Appendix). The executable file name is CADPAC. MAXIT and PRINT options set the maximum number of SCF iterations and set the amount of printout. The convergence option is set at 10^{-7}. Iteration stops when the root-mean-square density is not changed by more than 10^{-7} from its previous value.

TABLE 12-1 STO-3G basis function table (partial)

		Hydrogen (Scale Factor = 1.24)				
H1s	3.425250	0.623913	0.168858	0.154329	0.525328	0.444635
		Helium (Scale Factor = 1.69)				
He1s	6.362420	1.158923	0.313651	0.154329	0.535328	0.444635
		Lithium (Scale Factor = 2.69)				
Li1s	16.11957	2.936200	0.794654	0.154329	0.535328	0.444635
2s	0.444402	0.129955	0.042266	-0.099967	0.399513	0.700115
		Beryllium (Scale Factors = 3.68, 1.10)				
Be1s	30.16786	5.495113	1.487199	0.154329	0.535328	0.444635
2s	1.202986	0.279548	0.090918	-0.099967	0.399513	0.700116
		Boron (Scale Factors = 4.68, 1.45)				
B1s	48.79110	8.887359	2.405278	0.164329	0.535328	0.444635
2s	2.090312	0.485743	0.157980	0.099967	0.399513	0.700115
		Carbon (Scale Factors = 5.67, 1.72)				
C1s	71.61682	13.04509	3.530528	0.154329	0.535328	0.444635
2s	2.941250	0.683482	0.222291	-0.099967	0.399513	0.700115
2p	2.941250	0.683482	0.222291	0.155916	0.607684	0.391957
		Nitrogen (Scale Factors = 6.67, 1.95)				
N1s	99.10614	18.05231	4.885682	0.154329	0.535328	0.444635
2s	3.780457	0.878495	0.285716	-0.099967	0.399513	0.700115
2p	3.780457	0.878495	0.285716	0.155916	0.607684	0.391957
		Oxygen (Scale Factors = 7.66, 2.25)				
O1s	130.7093	23.80886	6.443637	0.154329	0.535328	0.444635
2s	5.033153	1.169594	0.380391	-0.099967	0.399513	0.700115
2p	5.033153	1.169594	0.380391	0.155916	0.607684	0.391957
		Fluorine (Scale Factors = 8.65, 2.55)				
F1s	166.6791	30.36080	8.216857	0.154329	0.535328	0.444635
2s	6.464805	1.502279	0.488591	-0.099967	0.399513	0.700115
2p	6.464805	1.502279	0.488591	0.155916	0.607684	0.391957

Below ATOMS in Fig. 12-1, comes the name *hydrogen* for the first atom and first atomic number, 1. The three floating-point numbers following the atomic number are the x, y, and z coordinates. Conveniently, this atom is placed at the origin, so these descriptors are all 0. The line 1 S 3 indicates that the 1s atomic orbital will be approximated by three primative gaussians.

Below this is a block of nine numbers taken from Table 12-1. The first is the number of the primative, the second in the same row is the orbital exponent, and the third is the contraction coefficient. (Actual numbers from different sources may differ slightly according to the fitting procedure, or they may differ widely if one set is for the normalized function and the other is for the nonnormalized function.) This is repeated for the second and third primatives to complete the block. END designates the end of the orbital definition for the first atom. Extensive tables of basis set functions have been published. Table 12-1 (from Hehre, Stewart, and Pople, 1969) is an example, containing the STO-3G basis set functions for hydrogen, the first row elements, and sulfur.

The second block of information for MICRO CADPAC is the same as the first except that, after designating the atomic number as 1.0, the position on the x axis is entered as 1.40 a.u. Because this is a single-point calculation, this value does not change. After an END designator terminates the second ATOM block, another END command shows that there are no more atoms to enter. Subprogram START is automatically loaded. Subprogram FINISH closes files and provides for an orderly FORTRAN termination.

Figure 12-2 shows a screen dump of messages sent to the CRT during the run. Most of these are self-explanatory in the light of the preceding information. The INITIAL GUESS is the first SCF matrix which, in this option, is the core hamiltonian HCORE. The final energy is given in double precision because the atomic unit is such a large unit of energy (1 a.u. = 2625 kJ mol^{-1}).

Results of the Calculation

After only two cycles of the calculation, this simple procedure terminates at a total energy of -1.1167 a.u. The exact value is estimated as -1.1744 a.u. (Parr, 1963). This result looks better than it is because of the size of the atomic unit. Basically it says that a chemical bond exists between H and H. The atomic energies are -1.0000 a.u., leaving -0.1167 a.u. = -306 kJ mol^{-1} as the bond energy of the H$_2$ molecule. Considerably more information on the calculation can be obtained by printing the data file OUTPUT, which is produced as the energy calculation is carried out.

EXERCISE 12-3

Carry out STO-3G calculations on H_2 at various interatomic distances from 0.8 to 2.6 a.u. of distance (Bohr). This is done by simply changing the entry 1.40 Bohr in the lower block of information in Fig. 12-1 to any desired internuclear distance for each new single-point calculation. Compare your results with the spoon-shaped curve of bonding energy as a function of internuclear distance given in elementary texts (e.g., Levine, 1988).

FIGURE 12-2 Screen dump of messages received from CADPAC during *ab initio* calculations on H_2.

```
 cadpac
DATIN LOADED
DATIN COMPLETED
INIT LOADED
INIT COMPLETED
START LOADED
START COMPLETED
INTEGRALS SECTION LOADED
CALCULATING ONE ELECTRON INTEGRALS
CALCULATING TWO ELECTRON INTEGRALS
INTEGRALS COMPLETED
SCF SECTION LOADED
INITIAL GUESS
HCORE


       RHF SCF CALCULATION


CONVERGENCE DATA
MAXIMUM NUMBER OF ITERATIONS=     20
CONVERGENCE CRITERION       = 1.0D - 7

CYC     TOTAL      ELECTRONIC    E CONV.   D CONV.   ICA DEL(T)  TIME   SHIF
        ENERGY      ENERGY
1    -1.1167144  -1.8310001  -1.8310001  .0000000    1  17.08  69.42   .00
2    -1.1167144  -1.8310001   .0000000   .0000000    2  10.00  79.42   .00


        FINAL VALUES OF TOTAL ENERGY AND ELECTRONIC ENERGY ARE
                     -1.116714389 AND       -1.831000104
START LOADED
START COMPLETED
RUN COMPLETED
Stop - Program Terminated.

c >
```

THE PROGRAMS

MICRO CADPAC (Appendix) runs in 256 K of memory under DOS 2.0 (note). It performs single-point closed-shell calculations on a maximum of 12 atoms with a maximum of 40 basis functions. Symmetry may be specified, but that option is not used in Computer Projects 12-1, 12-2, and 12-3. Orbital functions are not self-contained and must be specified in the input file (e.g., Fig. 12-1). Gradient or optimizing calculations are not carried out. Examples of STO-3G and 3-21G input files are given with the documentation. The advantage of this program is its simplicity. MICRO CADPAC is probably the smallest *ab initio* program available.

An expanded and updated version by the same authors called MICROMOL is available (Appendix). MICROMOL comes in two versions, one that runs in 640 K of active memory and has a maximum of 12 atoms and 63 basis functions. The second runs in 1 Mb of RAM (under OS/2) and has a capability of 30 atoms and 127 basis functions. Both optimize by the method of gradients. The larger has a library of basis functions and a graphics subpackage. We have used neither in our laboratory.

GAUSSIAN70 is program QCMPE 061 (Appendix). The package contains four parts, one for constructing an input file, one for checking the input file, the executable file G70.EXE, which carries out the *ab initio* calculation, and a file to print the results. The three satellite files, G70DATA, G70DATCH, and G70PRINT are useful mainly at the beginning: After a little experience, the user will probably use only G70DATA, and then only to build new files. Modification of files to solve a series of related calculations is best done using EDLIN. The executable file is large (about 330K) and fills one diskette. The program uses almost 640K of internal memory and requires several megabytes of hard-disk memory to store integrals. As many as 10 Mb of disk space may be needed for some problems, but the problems suggested in this chapter require only a fraction of that amount. The reader with a special interest in *ab initio* calculations may wish to dedicate a machine and hard disk entirely to this task.

The package received from QCPE is on 10 diskettes, the first two of which are the executable code for the four programs just mentioned. Executable code has been compiled for 80286 or 80386 microprocessors. If one wishes to run the program on an 8086 microprocessor, as we have done, it must be recompiled using MS FORTRAN 4.0. Compilation, linking, and installation are not trivial and are probably best done by an experienced programmer. All the necessary information is available in the MS FORTRAN manual and the GAUSSIAN70 documentation, however, and the diligent beginner can recompile, link, and install GAUSSIAN70.

Note that our compiler required blanks in the command FL/Os/G0/c G70*.FOR that are not as shown in the program documentation. We compiled all the G70 program modules in one run followed by PACKER.FOR in a second run and the satellite programs G70DATA.FOR, G70DATCH.FOR, and G70PRINT.FOR in another. Pay special attention to the directions on PATH, SET LIB, SET TMP, CONFIG.SYS, and AUTOEXEC.BAT.

GAUSSIAN70 carries out *ab initio* calculations on 30 atoms with up to 70 basis functions of the *s* and *p* type. Using the 8086 microprocessor, run times will be prohibitive for molecules of less than half this size. GAUSSIAN70 has a self-contained library of STO-nG and n-31G basis functions (typically, STO-3G and 4-31G). Optimization is carried out by the gradient method but only in a limited way. See Computer Project 12-3 for a discussion of optimization by GAUSSIAN70.

COMPUTER PROJECT 12-1 | *Ab Initio Energy and Wave Functions of HF*

Run CADPAC for single-point *ab initio* calculations on HF. Necessary input includes the atomic numbers of the atoms involved in the molecule (9 and 1), a starting bond distance, and the STO-3G orbital functions for fluorine and hydrogen taken from Table 12-1. Search the potential function for its minimum by varying the internuclear distance.

If GAUSSIAN70 is used, the orbital functions in Table 12-1 are contained in a library within the program and are not entered. The input file is in the form of a simple *z* matrix (Clark, 1985). A *z*-matrix input file is shown in Fig. 12-3. The package GAUSSIAN70 contains an interactive program G70DATA to help the beginner construct a *z*-matrix input file, which is stored under the name G70FILE5.DAT. Upon executing the *ab initio* program G70, whatever has been loaded into G70FILE5.DAT is used as input for the calculation. The output file is named G70FILE6.DAT

FIGURE 12-3 GAUSSIAN70 *z* matrix input file for HF.

```
     1   1   1   1   1   0   0
    HF, single point

    0 1
      9
      1   1 1.0000   0    .000000   0    .000000   0
          0
    -1
```

and may be examined at the CRT screen *via* a TYPE command or sent to the printer with CTRL P.

Procedure After the first successful run, it is not necessary to construct files anew. Small changes such as bond lengths can be made most easily using EDLIN. One may wish to start an archive of output files by copying G70FILE6.DAT to a unique file name.

Run as many single-point STO-3G calculations at different internuclear distances as are necessary to find the energy minimum. The experimental bond distance for HF is 0.917 Å (Malinson and Davies, 1978). Is the STO-3G internuclear distance the same as the experimental distance? Interpret the eigenvalues at the energy minimum in terms of s and p involvement in the bond.

COMPUTER PROJECT 12-2 | *The Dipole Moment of Water*

Perhaps the most striking molecular property of water is its dipole moment, which accounts for its ability to dissolve many ionic salts, its ability to hydrogen bond, and, in part, its ability to sustain life. In this project, we shall use an STO-3G basis set to carry out *ab initio* calculations of the dipole moment of water. Using CADPAC, this requires three blocks of information in the input file: two for hydrogen and one for oxygen. The usual way of doing this is to place oxygen at the origin and to indicate the angles between the two hydrogens.

Using GAUSSIAN70, only the atomic numbers of the constituent atoms, two bond lengths (which are the same), and a bond angle need be input, along with some control information. An appropriate G70FILE5.DAT input file is shown in Fig. 12-4. The first line consists of a series of option switches that, in essence, specify an STO-3G, single-point calculation. Other options in this line will become self-evident upon using G70DATA. The second line is a comment line and the third specifies the uncharged ground state (also self-evident from G70DATA). The next three

FIGURE 12-4 GAUSSIAN70 z matrix input file for H_2O.

```
    1   1   1   1   1   0   0
   water sto-3G geometry (hehre et al. p138)

   0 1
     1
     8   1  .9900   0     .000000   0   .000000   0
     1   2  .9900   1 100.000000   0   .000000   0
         0
    -1
```

lines are the atomic number–geometry block or z matrix. Column 1 contains the atomic numbers of H, O, and H. Row 2 specifies that atom 2, oxygen, is attached to atom 1 at a distance of 0.990 Å. Row 3 stipulates that atom 3 is attached to atom 2 at a distance of 0.990 Å and makes an angle with atom 1 of 100°. Lines 7 and 8 are switches designating the end of the z matrix and the end of the file, respectively.

Procedure

(a) Determine the dipole moment of water for the geometry given in Fig. 12-4. This is the geometry from a previous optimization using the STO-3G basis set with a mainframe computer. Compare your result with an experimental value from the literature.

(b) Determine the dipole moment of water by a single-point calculation using the experimental values of the bond length and bond angle, 0.958 Å and 104.5°, respectively.

(c) At constant bond length carry out repeated calculations of the dipole moment for bond angles from 10 to 180° at intervals of 10°. Plot the function of dipole moment μ vs. bond angle θ over the entire interval.

COMPUTER PROJECT 12-3 | *Gradient Calculations*

By combining the concepts of *ab initio* calculations with the minimum search technique of molecular mechanics, one arrives at the method of *gradient ab initio* calculations. In this method, *ab initio* calculations are carried out, followed by a search of the potential energy surface as a function of small variations in bond lengths and angles. The direction of steepest descent on the surface—its largest energy gradient—is selected, and bond lengths and angles are changed so as to seek a minimum on the surface in the way that is done in a molecular mechanics calculation. The *ab initio* calculation is carried out again and a new energy and energy gradient are found.

The entire process is repeated until the energy difference between two calculations has been reduced to some predetermined limit. The final geometry is said to have been *optimized* with respect to geometry and represents the best molecular parameters that can be obtained from that basis set. We shall use the minimum or STO-3G basis set in this exercise. The results will not be at the Hartree–Fock limit, but the geometries are usually rather good, even with this small basis set.

CADPAC does not carry out gradient calculations, although a minimum can be found for simple molecules by repeated single-point calculations as in Computer Project 12-1. We shall optimize the geometry for hydrogen and methane at the STO-3G level using GAUSSIAN70.

GAUSSIAN70 does not carry out a full energy minimization either, but it approximates the optimum geometry by a three-step process. The

```
    1   2   1   1   1   0   0
hydrogen, gradient

  0 1
    1
    1   1   .9000   0   .000000   0   .000000   0
        0
    1                           .10
    0
   -1


    1   2   1   1   1   0   0
hydrogen, gradient

  0 1
    1
    1   1   .8000   0   .000000   0   .000000   0
        0
    1                           .01
    0
   -1
```

FIGURE 12-5 (a) Initial and (b) modified input files for gradient *ab initio* calculations on H_2.

energy is calculated for the input geometry, for the original geometry varied by a small amount, and then varied again to obtain a third geometry. The three points are used to fit a parabola (Chapter 5). The minimum of the parabola is taken as the energy at the geometric parameter in question. For example, if the bond distance for H_2 is guessed as 0.9 Å and the variation parameter is set at 0.10 Å, the input data file in Fig. 12-5 can be generated using the interactive program G70DATA. Upon running G70, the first approximation to the optimized bond length of hydrogen is 0.574 Å. This is not very good but it shows that the bond length is less than 0.9 Å and that the variation parameter 0.1 Å is too gross.

Using EDLIN, G70FILE5.DAT can be altered to Fig. 12-5(b), in which the starting geometry has been changed to 0.8 Å and the variation parameter or step size for the parabolic extrapolation has been changed to 0.01 Å (the smallest step size permitted). The extrapolated bond length is now 0.685 Å, which is a better approximation to the experimental value of 0.74144, but is still poor. Optimization is not complete, even at the STO-3G level.

Procedure

(a) Continue optimization by using 0.685 Å as the starting bond length of H_2. Repeat, using the new bond length. Repeat until successive bond lengths are self-consistent to within 0.001 Å. Compare your result with the experimental value.

(b) Assuming tetrahedral geometry, optimize the bond lengths of the C—H bonds in methane. Start from an approximate bond length of 1.0 Å and optimize in the STO-3G basis set. Placing one hydrogen atom at the origin makes all simple angles 109.47°.

(c) Repeat the minimization of part (a) to obtain the bond length of LiH. Plot the bond lengths of the hydrides of H, F, O (Computer Projects 12-1 and 12-2), and C (methane). Interpolate the predicted bond lengths of covalent hydrides of B and Be. Compare your values with a tabulation of experimental bond lengths (Sanderson, 1976).

PROBLEMS | Chapter 12

1 The Schroedinger equation for the helium atom is

$$\left[-\frac{1}{2}\nabla_1^2 - \frac{1}{2}\nabla_2^2 - \frac{Z}{r_1} - \frac{Z}{r_2} + \frac{1}{r_{12}} \right]\psi = E\psi$$

If we completely neglect the $1/r_{12}$ term, this is simply the sum of two hydrogen-atom equations. Using this crude first approximation, write the total energy of the helium atom in atomic units of hartrees, in electron volts, and in kilojoules per mole.

2 To improve the approximation in Problem 1, we can solve the energy integral

$$E = \langle 1s | H | 1s \rangle$$

using the hydrogen $1s$ wave functions for helium but with $Z = 2.0$. This is still an approximation because the wave function for hydrogen is not the same as the wave function for helium. The integral can be evaluated (McQuarrie, 1983) and yields

$$E = Z^2 - \left(\tfrac{27}{8}\right)Z$$

Calculate the energy. Using the criterion of energy minimization, is this a better or worse approximation than the one made in Problem 1?

3 It is reasonable to suppose that the effective nuclear charge Z' is less than 2.0 because the negative charge on one electron "shields" the full nuclear charge from the other. If the effective nuclear charge is 0, however, the energy is also 0, which is greater than the energies calculated in both Problems 1 and 2. There must be an effective nuclear charge between 0 and 2.0 that minimizes the energy. Write a short program in BASIC to find E and Z at the minimum of $E = f(Z)$. Check your results by setting $(dE/dZ) = 0$ for the expression in Problem 2 and solving analytically.

4 We know the exact solution for the helium ion He^+: It is $-Z^2/2.0$ hartrees $= -54.42$ eV. There is no second electron; hence, no shielding and $Z = Z' = 2.0$ for the ion. We are now in a position to calculate the ionization potential of helium:

$$He \longrightarrow He^+ + e^-$$

It is simply the difference between the approximate value of the energy of He and the exact value of the energy of its ion He^+. Calculate the ionization potentials for each of the three approximations in Problems 1, 2, and 3. Compare the results with the experimental value for the ionization energy of helium, which is 24.6 eV.

5 What is the energy of the σ_u orbital in the H_2^+ ion, at an internuclear separation of 1.40 a.u., in units of kilojoules per mole? Is this a bonding or an antibonding orbital?

6 Consult the literature (Dewar, 1971) for the LCAO calculation of bonding energy in H_2 as a function of internuclear distance. Plot $E = f(R)$ on the same graph with the results of Exercise 12-3 for comparison. How do the curve shapes and minimum energies compare? How do the minimum energies (LCAO and HF–SCF) compare with the experimental energy of -465 kJ mol^{-1}? How does this compare with the value of the dissociation energy of H_2 that you obtained in Computer Project 2-6?

7 Using STO-3G, determine the bond length of the diatomic molecules Li_2 and LiF. Which of these do you expect to coincide more closely with the experimental value and why?

8 Using GAUSSIAN70 (QCMPE 061), repeat the single-point calculation on the equilibrium bond energy of H_2 that was carried out with CADPAC in this chapter. Note that the bond distance is specified in angstroms in GAUSSIAN70; hence, $BL = 1.40$ a.u. $= 0.739$ Å.

9 Repeat Exercise 12-3 using GAUSSIAN70. Are the results identical to those obtained with CADPAC?

10 If the water molecule were stretched so that the two O—H bonds were exactly 1.0 Å and the angle between them was 105.0°, what would its dipole moment be?

11 Using Pythagoras' theorem, calculate the simple angle H—C—H of a regular tetrahedron where H represents an apex and C represents the point at the center.

12 Determine the bond lengths and angles of ammonia (NH_3) in the STO-3G basis set.

REFERENCES

Clark, T., 1985. *A Handbook of Computational Chemistry*. Wiley, New York.

Coulson, C. A., 1937. *Trans. Faraday Soc.*, **33**, 1479.

Dewar, M. J. S., 1971. *J. Chem. Ed.*, **48**, 494.

Hehre, W. J., Radom, L., Schleyer, P. v. R., and Pople, J. A., 1986. *Ab Initio Molecular Orbital Theory*. Wiley, New York.

Hehre, W. J., Stewart, R. F., and Pople, J. A., 1969. *J. Chem. Phys.*, **51**, 2657.

Hinchliffe, A., 1988. *Computational Quantum Chemistry*. Wiley, New York, pp. 106–108.

Huzinaga, S., 1984. *Basis Sets for Molecular Orbital Calculations*. Elsevier, New York.

Klopman, G. and Evans, R. C., 1977. *Modern Theoretical Chemistry*, G. A. Segal, ed. Plenum, New York, Vol. 7, Chapter 2.

Levine, I. N., 1988. *Physical Chemistry*, 3rd ed. McGraw-Hill, New York, Figure 20.2.

Malinson and Davies, P., 1978. *Chemistry in Britain*, **14**, 191.

Mc Quarrie, D. A., 1983. *Quantum Chemistry*. University Science Books, Mill Valley, CA.

Parr, R. G., 1963. *Quantum Theory of Molecular Electronic Structure*. Benjamin, New York, Table 2.

Richards, W. G. and Cooper, D. L., 1983. *Ab Initio Molecular Orbital Calculations for Chemists*. Oxford, New York.

13

Advanced Semiempirical Molecular Orbital Methods: MNDO

Several advanced semiempirical SCF–MO methods are in use. All entail neglect of interactions between electrons that are not in the valence shell. In *ab initio* theories, the molecular core is defined as the nuclei only (Chapter 12). Here, however, we define the core as the nuclei plus the associated nonvalence electrons (Dewar and Thiel, 1977). The core influences valence electrons through its potential energy field.

Among valence electrons, there is some degree of neglect of differential overlap (NDO), that is, some overlap integrals are arbitrarily set equal to zero (Klopman and Evans, 1977). Coulomb and exchange energies arising from orbitals that are not neglected contribute to the total electronic energy. In semiempirical MO models, these integrals are not evaluated; their energies are approximated by empirical parameters. Thus, of the millions of integrals that are generated in a full *ab initio* treatment of an organic molecule, many disappear because nonvalence electrons are lumped together in the core, many are neglected by setting them equal to zero, and the remainder are approximated by fitted parameters.

The logic followed in this chapter will be to develop the Fock matrix as in Chapter 12, to describe the approximations made, and then to discuss the parameter fitting procedures in general terms.

DIFFERENT SEMIEMPIRICAL METHODS

NDO approximation schemes (CNDO, INDO, MINDO, etc.) differ in the point, during the logical development, at which approximations are made and in the infusion of empirical data that takes place. In early theories, elements of the Fock matrix were set equal to zero or to one of a small number of empirical parameters (Chapter 9). In more recent methods, the logic of the *ab initio* method is followed as far as the Roothaan equations, whence restrictions are placed on the overlap matrix such that its off-diagonal elements are set to zero [the neglect of diatomic differential overlap (NDDO) approximation]. Remaining elements in the Fock matrix are parametrized and the normal equations are brought to self-consistency by the SCF method.

Parametrization entails selecting a small number of empirical parameters that can be used in an equation or equation set to calculate a subset of the physical and chemical properties of a select group of molecules. Calculated values of the properties in question are matched against experimental measurements and the differences are minimized, usually by some variant of the least squares procedure (Chapters 5 and 6), to obtain the parameter set that best reproduces experimental results.

Parametrization forces a certain amount of selectivity in the category of problem best solved by a semiempirical method (Chapter 9). Methods parametrized on molecular spectra give good results for other molecular spectra, methods parametrized on dissociation energies give good dissociation energies, and so on. Parametrization may be very extensive in the more recent semiempirical methods; some parameter sets have more than 100 members.

Dewar's NDO programs constitute a series evolved to give good ionization and dissociation energies. They were not developed, as some earlier programs were, to serve as a cross-check on *ab initio* calculations. They are used to generate chemically useful information on molecules that are not presently in the range of *ab initio* calculations. One of the most successful of this series is program MNDO (Dewar and Thiel, 1977). This is the method that we shall use in the computer projects.

MNDO Initial Approximations

Several fundamental approximations are made at the outset. The treatment is confined to closed-shell molecules, that is, molecules in which all electrons are paired. The nuclei and all nonvalence electrons are regarded as the *core*, which influences valence electrons through its charge but does not enter into chemical bonding. The number of integrals is restricted by taking a minimum basis set of valence shell atomic orbitals and by neglecting integrals involving diatomic differential overlap (Dewar and Thiel, 1977).

The equation set used as the starting point of the MNDO method is the Roothaan equation set [Chapter 12, Eq. (12-36)], which has the normal equations [Eq. (12-39)]

$$\sum \left(F_{\mu\nu} - E_i S_{\mu\nu} \right) a_{\nu i} = 0 \tag{12-39}$$

These are normal equations in the unknowns $a_{\nu i}$ and contain the orbital energies E_i. The operator in the general Roothaan equation set is the Fock operator $\mathbf{F}$ of *ab initio* theory. The symbol $\mathbf{H}$ is retained, but will be used from now on to designate the hamiltonian operator $\mathbf{H}^{core}$ that generates the energy of an electron moving in the potential field due to all nuclei and the inner shell electrons of the molecule.

S is the matrix of the overlap integrals with elements

$$S_{\mu\nu} = \langle \phi_\mu | \phi_\nu \rangle$$

If we stipulate that the overlap matrix must be equal to the identity matrix (Chapter 3), we are saying, physically, that the only overlaps we shall consider are those between orbitals on the same atom. The Roothaan equations become

$$\sum (F_{\mu\nu} - E_i \delta_{\mu\nu}) a_{\nu i} = 0 \qquad (13\text{-}1)$$

where $\delta_{\mu\nu}$ is the Kroenecker delta, equal to 1 for $\mu = \nu$ and equal to 0 for $\mu \neq \nu$. This is the NDDO (neglect of differential diatomic overlap) approximation. The electronic energy is given by a one-electron contribution:

$$E_{el} = \tfrac{1}{2} \sum \sum P_{\mu\nu}(H_{\mu\nu} + F_{\mu\nu}) \qquad (13\text{-}2)$$

where $H_{\mu\nu}$ is the core hamiltonian and $P_{\mu\nu}$ is an element in the density matrix. Now take atomic orbitals ϕ_μ and ϕ_ν to be centered on atom A and ϕ_λ and ϕ_σ to be centered on atom B. The general or *ab initio* Fock matrix elements are (Chapter 12)

$$F_{\mu\nu} = H_{\mu\nu}^{\text{core}} + \sum \sum P_{\lambda\sigma}[\langle \mu\nu | \lambda\sigma \rangle - \tfrac{1}{2}\langle \mu\lambda | \nu\sigma \rangle] \quad (12\text{-}42)$$

where

$$H_{\mu\nu}^{\text{core}} = \langle \phi_\mu(1) | \mathbf{H}^{\text{core}} | \phi_\nu(1) \rangle$$

It is the two-electron repulsion integrals $\langle \mu\lambda | \nu\sigma \rangle$ that cause most of the trouble and that are neglected in NDDO theory. A rationale for neglect of differential overlap approximation can be found in Dewar's summary of semiempirical MO theory (Dewar, 1975).

MNDO—Basic Theory

Under the NDDO approximations, the diagonal elements of the Fock matrix become

$$F_{\mu\mu} = U_{\mu\mu} + \sum_B V_{\mu\mu, B} + \sum_\nu^A P_{\nu\nu}[\langle \mu\mu | \nu\nu \rangle - \tfrac{1}{2}\langle \mu\nu | \mu\nu \rangle]$$

$$+ \sum_B \sum_{\lambda, \sigma}^B P_{\lambda\sigma}\langle \mu\mu | \lambda\sigma \rangle \qquad (13\text{-}3)$$

where $U_{\mu\mu}$ is the one-electron one-center electronic energy, that is, the kinetic energy of an electron in orbital ϕ_μ of atom A plus the potential energy exerted on it by the core at atom A. $V_{\mu\mu, B}$ is a one-electron two-center potential exerted by the core at B on an electron in the orbital ϕ_μ at atom A. $P_{\nu\nu}$ is a density matrix element, $\langle \mu\mu | \nu\nu \rangle$ are one-center coulomb integrals, and $\langle \mu\nu | \mu\nu \rangle$ are one-center exchange integrals. The integral $\langle \mu\mu|\lambda\sigma \rangle$ is a two-center two-electron repulsion integral between an electron in orbital ϕ_μ of A and an electron in distribution $\psi_\lambda\psi_\sigma$ on atom B.

The off-diagonal elements for ϕ_μ and ϕ_ν on the same atom are

$$F_{\mu\nu} = \sum_B V_{\mu\nu, B} + \tfrac{1}{2}P_{\mu\nu}[3\langle \mu\nu|\mu\nu \rangle - \langle \mu\mu|\nu\nu \rangle]$$

$$+ \sum_B \sum_{\lambda,\sigma}^B P_{\lambda\sigma}\langle \mu\nu|\lambda\sigma \rangle \tag{13-4}$$

where $V_{\mu\nu}$ is a two-center one-electron attraction between the core of B and an electron in the distribution $\psi_\mu\psi_\nu$ of atom A. The integrals $\langle \mu\nu|\lambda\sigma \rangle$ are two-center two-electron repulsion integrals.

The remaining equation for off-diagonal elements with ϕ_μ and ϕ_λ on different atoms is

$$F_{\mu\lambda} = \beta_{\mu\lambda} - \tfrac{1}{2}\sum_\nu^A \sum_\sigma^B P_{\nu\sigma}\langle \mu\nu|\lambda\sigma \rangle \tag{13-5}$$

It has β as a two-center one-electron core resonance integral. The notation β is intentional; note the analogy between β of Eq. (13-5) and β of SCF theory (Chapter 9). Equations (13-3) through (13-5) are given as Eqs. (4) through (6) in the original publication (Dewar and Thiel, 1977). These equations are notationally more complicated than the general equation [Eq. (12-42)], but they are easier to evaluate because they lack the two-electron two-center integrals $\langle \mu\lambda|\nu\sigma \rangle$.

Parametrization

One-center terms, $U_{\mu\mu}$, $\langle \mu\mu, \nu\nu \rangle$, and $\langle \mu\nu, \mu\nu \rangle$, are approximated by fitting to spectroscopic data. Empirical evaluation of these terms brings some degree of compensation for electron correlation. For this reason, both $\langle \mu\mu|\nu\nu \rangle$ and $\langle \mu\nu|\mu\nu \rangle$ are smaller than *ab initio* values without correlation.

Two-center repulsion integrals represent the energy of interaction between the charge distribution $e\phi_\mu\phi_\nu$ of atom A and $e\phi_\lambda\phi_\sigma$ of atom B, where e is the charge on the electron. Based on a classical model of charge interaction, Dewar has obtained a semiempirical function of distance $f_1(R_{ij})$ between point charges i and j, where the distance R_{ij} is determined from the internuclear distance R_{AB} by a functional relationship fitted to give correct values in the limits of $R_{AB} = 0$ and ∞. Comparison of the semiempirical function described by Dewar with the analytical integrals shows that these functions are also smaller, due to partial compensation for electron correlation effects.

Core–electron attraction $V_{\mu\nu, B}$ and core–core repulsions E_{AB} are given by two more expressions containing empirical functions $f_2(R_{AB})$ and $f_3(R_{AB})$ of internuclear distance. Like the electron repulsion function f_1, f_2, and f_3 are approximated by a simplified classical model of electrostatic attraction or repulsion of valence-shell charge distributions for the core.

The one-electron resonance integrals β are taken to be proportional to the overlap integrals through another empirical function

$$\beta_{\mu\lambda} = f_4(R_{AB})S_{\mu\lambda} \qquad (13\text{-}6)$$

These are the integrals of greatest importance in chemical bonding. As in Chapters 7 through 9, bonding is proportional to overlap S.

A nonlinear least squares procedure yielded the best values for f_1 through f_4. As it happens, f_2 is very small and can be replaced by 0, yielding a purely classical expression for the core–electron attraction energy.

The total energy is calculated as the sum of valence electronic energy and the repulsions between the cores summed over all atom pairs:

$$E = E_{el} + \sum_{A < B}\sum E_{AB}^{core} \qquad (13\text{-}7)$$

The enthalpy of formation is

$$\Delta H_f = E - \sum_A E_{el}^A + \sum_A \Delta H_f^A \qquad (13\text{-}8)$$

(Bingham, Dewar and Lo, 1975) where E_{el}^A is the electronic energy and ΔH_f^A is the heat of formation of each atom *in the molecule*.

Once enough fitting parameters have been obtained to generate an approximate Fock matrix, repeated diagonalization leads to self-consistency as in the PPP method.

In MNDO, for H, C, N, and O, there are about 30 optimized and about 30 derived parameters. MNDO contains only atom parameters whereas MINDO/3, which Dewar feels is superceded by MNDO, contains both atom and bond parameters, and more of them. The discrepancy between MNDO calculated values and experimental values for most molecular geometries and energies is one-half to two-thirds that of the MINDO/3 values.

THE PROGRAM

The program used in the exercises is a modification of Thiel's original MNDO program (Dewar and Thiel, 1977). The original program was issued for mainframe as QCPE 353. The modification for microcomputer (IBM-PC and compatible), including source code, executable code, and documentation, is available from Serena Software (Appendix). Parameters in the version used here were updated 3/86.

The latest development in this series is AM1, which is available in mainframe version but not, at this writing, in microcomputer format. The mainframe version is part of MOPAC (a general molecular orbital package, QCPE 455), which is in the public domain. MOPAC contains both MNDO and AM1 and permits the operator to select either as an option, using keywords at the beginning of the run. The keyword AM1 selects the later method. MNDO is the default method.

Input Files

Input files are very similar to the standard z matrix described in the *Handbook of Computational Chemistry* (Clark, 1985). Details of the input files can be obtained directly from the MNDO program disk in the form of a document file called **mndo.doc**, which gives a simplified set of rules and recommendations to get started, along with a detailed list of options and defaults in the program. The example and computer projects that follow were selected because they require simple input files and run in reasonable time.

The z matrix is an array of numbers that gives the geometry of the target structure, along with some options, such as the option to optimize a bond length or angle, or the option to leave it frozen as in a single-point *ab initio* calculation. Efforts have been made to standardize the z-matrix format and z matrices are similar, even in calculations that are based on very different premises, such as GAUSSIAN80 and MNDO. Minor differences in z-matrix format still exist, however. Contrast the z matrices in Clark (1985) with the z matrices used in the Serena Software MNDO program.

Dummy atoms are used to simplify input of the starting geometry. Thus, suppose we wish to enter an atom and then place a second atom at a certain distance from it, but at an angle to the x axis, in preparation for placing a third atom at some angle to the x axis, and so on. We place a dummy atom at the origin, the first atom at some arbitrary distance from the dummy, and the second atom at the desired distance from the first and at the desired angle to the line established by the positions of the dummy and the first atom (the x axis). In this way, triatomic molecules, for example, can be entered without using negative coordinates.

The Methylene Radical (Clark, 1985)

We shall construct the input file named MET.DAT (capitals not necessary) for the singlet methylene radical as an example (Fig. 13-1, top). Formats are in parentheses.

```
         met.dat

99
 6       1.0                                                           1
 1       1.09        1         122.0       1         0.0       0       2 1 0
 1       1.09        0         122.0       0       180.00      0       2 1 3
99



         h2.dat


 1       0.0                                                           0
 1       1.40        1          0.0       0         0.0       0       0 0 0
99



         ethane.dat

 0

 6       0.0                                                           1
 6       1.54        1                               0               2 1
 1       1.09        0         110.0       0                 0       2 1 3
 1       1.09        0         110.0       0       120.00      0       2 1 3
 1       1.09        0         110.0       0       240.0       0       1 2 3
 1       1.09        0         110.0       0        60.00      0       1 2 3
 1       1.09        0         110.0       0       180.00      0       1 2 3
 1       1.09        0         110.0       0       300.0       0       1 2 3
 0
99
```

FIGURE 13-1 MNDO z-matrix input files for methylene, H_2, and ethane.

Line Entry
1 Blank
2 Zero (I2) or blank Sets the maximum number of SCF iterations at the default limit of 9999.
3 Blank
4 99 (I2) Specifies an arbitrary geometric reference point. Sometimes called a *dummy atom* #1. (There is no real atom at this position.)
5 First real atom position. 06 (I2) designates the identity of atom 2 by its atomic number (6: carbon). The entry 1.0 (F10.5) gives the distance in angstroms from dummy atom 1. The 1 at the end of the line (I2 in columns 71–72) specifies that the distance is to be measured relative to atom 1 (the dummy).
6 Hydrogen (atomic number 1) is 1.09 Å from atom A, designated below. The bond length is to be optimized (1 in I2 format in columns 23–24). It

makes an angle of 122.0° (also to be optimized) with the line from atom 1 (dummy) and atom 2 (carbon). There is no dihedral angle to worry about. Hence, call it 0.0 and do not optimize (0 in 63–64). The angle *A-B-C* is designated by the three numbers in I2 format 2 1 0 at the end of the line.
7 The second hydrogen is 1.09 Å from atom *A*, makes a 122.0° angle with the line *AB* and has a 180° dihedral angle measured clockwise *C-B-A–This atom* measured about the axis from *B* to *A*. No angle or distance is optimized, as indicated by the three 0 switches in this row. Atoms *A*, *B*, and *C* are designated 2, 1, and 3 by the three digits at the end of the row.
8 Zero or blank indicates the end of the geometry designators.
9 99 indicates that there are no data sets following this one.

The entire input file is shown at the top of Fig. 13-1. After installation of the MNDO program (instructions are given on the disk), typing the executable filename MNDO produces a short introductory screen with an input prompt at the bottom. Respond with met.dat (or any other named input file). If a FORTRAN stop appears in a few seconds, the run has failed and the input file should be checked for errors. If the run takes about a minute, it is successful. Messages are sent to the CRT screen at intermediate stages of the program run to indicate its progress. At the conclusion of the run, a heat of formation is printed on the CRT screen. If this value is within 0.001 kcal mol^{-1} of the mainframe MNDO value of 107.366 kcal mol^{-1}, the first demonstration is complete.

The first demonstration approximates a full mainframe run but is not identical to it. Although the energy is close, some geometric parameters are poor. One can easily see the reason by typing TYPE MET.MNO. This produces the full output file to the CRT screen. The output can also be routed to a printer by pressing CTRL PRINT (simultaneous depression of the control and print keys) previous to the TYPE command. Upon inspection, one sees that the two C—H bond lengths are different, and that the angles made with the *x* axis are different. This is because the optimize switches were set to 1 for one hydrogen and to zero for the other. One hydrogen was optimized and the other was frozen.

The correct output file can be obtained by setting all length and simple angle optimization switches to 1. The dihedral optimization switches are still zero because a triatomic molecule is planar. Run the full optimization. This constitutes demonstration two and generates the correct MNDO output file. An archive of input and output files can be kept simply by renaming the input file before minimization. For example, COPY MET.DAT MET1.DAT before changing the optimization switches retains both files for future reference. Running the two input files produces two complete output files MET.MNO and MET1.MNO for archival purposes. Other interesting files are produced, for example, type the MET1.INP file after the MET1 run and compare it with MET1.DAT.

Symmetry

The method of complete optimization is time-consuming. Run time for demonstration 2 is about 20% longer than demonstration 1. The situation becomes worse for larger molecules. There is a quicker way to achieve the same purpose that was achieved in demonstration 2.

Noting the plane of symmetry through the carbon atom, we can optimize one hydrogen atom, then place the second hydrogen in a symmetrical location below the plane of symmetry. This is indicated in the input file by going back to MET.DAT (the faster of the two runs). Copy to MET2.DAT and make the following changes:

1. Place a 1 in column 22 of line 3. This is a switch that says "symmetry conditions will be read in."
2. Enter the matrix

$$\begin{pmatrix} 3 & 1 & 1 & 4 \\ 3 & 2 & 1 & 4 \end{pmatrix}$$

using the format I2, 2X, I1, 5X, I2, 8X, (I2, 3X, ...). See **mndo.doc** on the MNDO disk for further details.

The first column indicates the atom from which symmetry will be taken. Column 2 indicates the kind of symmetry: 1 for bond length and 2 for angle with the x axis. Column 3 indicates how many atoms are to be symmetrically placed and column 4 indicates the atom to be located by symmetry. Thus, the first row of the matrix says "make the bond length of atom 4 the same as that of atom 3." The second row of the matrix says the same of the simple angle. A run on the symmetry-restricted input file produces the accuracy of MET1 at the speed of MET.

COMPUTER PROJECT 13-1 | Homonuclear Diatomic Molecules

Because there are no angles to worry about, the MNDO input files for homonuclear diatomic molecules are very simple. One need only specify the atomic number of the constituent atoms and make a starting guess at the bond length. Molecular hydrogen, for example, requires a first line of 1 for the atomic number of one hydrogen atom and a second line of 1 for the atomic number of the second atom, followed by a 0.90, which is our guess for the bond length of H_2. The 1 following the initial guess for the bond length indicates that the geometry is to be optimized. Without it, the bond length would be frozen at the initial guess.

MNDO does not accept a frozen bond length in a diatomic molecule; it requires at least one variable in the geometry. Larger molecules can be frozen in all but one of their geometric variables. Optimization switches are a useful feature in studying nonequilibrium geometries that do not reside at potential energy minima; see Computer Project 13-3.

In this program and in contrast to the *ab initio* program, the bond length is in angstroms. The format of the input line for this program is I2, F10.5, I2; that is, the first entry of each line must be right justified in the first two columns, the second entry must be in the following 10 columns, and the entry following that must be right justified in the next two columns.

1

1   0.90   1

99

Note that three blank lines precede the first data entry line and one blank line separates the last data entry from the 99 designating the end of the input file.

Interpretation of Results The heat of formation of any element in the standard state should be zero. A nonzero value for the enthalpy of formation of an element is to be expected from an approximate calculation as an artifact or error of the calculation. Compare the experimental ionization potential with the experimental value of 15.98 eV (Turner et al., 1970). Compare the optimized bond length with the experimental value of 0.741 Å (Hertzberg, 1970). See also, Dewar and Thiel (1977).

COMPUTER PROJECT 13-2 | *Prediction of Photoelectron Spectra from Orbital Energies by MNDO*

When high-energy radiation strikes a molecule, a binding electron of the molecule may be struck so hard by the incoming photon that it is dislodged, leaving behind a molecular ion

$$M + h\nu \rightarrow M^+ + e^-$$

An ionization energy of the molecule I can be determined from the principle of conservation of energy

$$h\nu = \tfrac{1}{2}mv^2 + I$$

where $\tfrac{1}{2}mv^2$, the kinetic energy of the ionized electron, can be measured with an electrostatic analyzer (Atkins, 1986). Because binding electrons

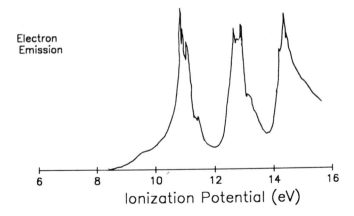

Electron
Emission

Ionization Potential (eV)

FIGURE 13-2 Calculated and observed photoelectron spectra of ethene.

occupy different orbitals, their ionization energies are different. A spectrum of peaks is obtained, each at or clustered about a different value of I, one I value being characteristic of ionization of an electron from each occupied orbital.

Procedure

(a) The photoelectron spectrum of ethene is sketched in Fig. 13-2. It shows several sharp peaks that are identified with the ionization potentials of electrons from the various orbitals in ethene. Redraw the PES on a piece of graph paper and superimpose vertical lines on this graph at each calculated value of molecular orbital energies. That is, if there is an eigenvalue at 10.5 eV, draw a vertical line there. From proximities of PES peaks and eigenvalues, identify the molecular orbital energy level that is responsible for each peak in the PES. Identify the degenerate levels, if any. Discuss the correspondences that you find. Why are some orbitals apparently not represented in the PES?

(b) Carry out both MNDO and GAUSSIAN70 calculations on N_2. Identify the orbitals corresponding to the PES spectrum of N_2. Draw the energy-level manifold of N_2 (not necessarily to scale). Identify the degenerate peaks. What is the first ionization potential of nitrogen gas? Does the N_2^+ ion have a stronger or weaker bond than N_2? Discuss this answer in the light of your *ab initio* and MNDO results.

COMPUTER PROJECT 13-3| *Isomeric Reactions*

All advanced MO calculations are approximate at the level of chemical accuracy. By selecting reactions in which the reactant is quite similar to the product, errors can be made to cancel and nearly chemical accuracy can be obtained. One class of reactions that often gives results at the level of chemical accuracy is isomerization. There are the same number of atoms on both sides of the reaction and most of the bonds are of the same kind. The isomerization of propene to cyclopropene is a case in point

$$CH_3CH{=}CH_2 \longrightarrow \underset{\underset{CH_2}{\diagup\diagdown}}{CH_2{-}CH_2}$$

Calculate the isomerization enthalpy for this reaction using MNDO.

COMPUTER PROJECT 13-4 | *The Enthalpy of Hydrogenation of Ethene*

The input file for ethane is similar to that of ethene but has, of course, more hydrogens. Start with all geometries save one frozen at reasonable angles and lengths as shown at the bottom of Fig. 13-1. Relax each atom stepwise by replacing the zero switch with a 1 switch for optimization. Do this line by line, yielding six separate calculations, one for each atom relaxed. Plot the results for each calculation as a function of the number of atoms relaxed, one for the calculation proceeding from Fig. 13-1, the second for the second line optimized, etc. What is the relaxation enthalpy per atom? What is the final and best value for the enthalpy of formation of ethane? Repeat for ethene. What is the best enthalpy of formation of ethene? What is the calculated enthalpy of hydrogenation of ethene to ethane? Compare this with the experimental value of -32.6 kcal mol^{-1} (Kistiakowsky, 1935).

COMPUTER PROJECT 13-5 | *The Rotational Barrier of Ethane*

Determine the peak of the rotational barrier of ethane by running MNDO with the dihedral angles of both methyl groups frozen at the same values. This conformation represents the eclipsed form of ethane. The enthalpy of formation will not be the same as the optimized molecule because optimization brings about relaxation to the staggered conformer, which is at the potential energy minimum. The barrier height is the difference between the enthalpies of formation of the two conformers.

Repeat the calculation with the dihedral or twist angles frozen at 10° difference. Repeat at 20, 30, 40, 50, and 60° differences. What do you find at 60° difference? Plot the energy difference between the frozen conformer and the relaxed conformer (Computer Project 13-2) as a function of the angular difference between hydrogens of the two methyl groups. By symmetry, the energy for 10° differences is the same as $-10°$, etc.; hence, you can make a plot extending from -60 through 0 to $+60°$.

PROBLEMS | *Chapter 13*

1 Modify the input file for H_2 so as to determine the bond length and ionization potential of N_2. Compare your calculated values with experimental results.

2 It is a simple step from homonuclear diatomic molecules to heteronuclear diatomics. Calculate the bond length, ionization potential, and *dipole moment* of carbon monoxide (CO).

3 The PES spectrum of N_2 was investigated using He(I) radiation (energy: 21.22 eV). Photoelectrons with a kinetic energy of 5.63 eV were detected. What was the ionization potential of these electrons?

4 Use the results of Problem 1 to determine the orbital from which the 5.63-eV photoelectrons of N_2 were ejected.

5 HBr has electrons of two kinds, bonding electrons and nonbonding $2p$ electrons of Br. What is the approximate ionization potential of each kind? Which is more firmly held by the HBr molecule?

6 Determine the enthalpy of isomerization of *cis*-buta-1,3-diene to *trans*-buta-1,3-diene by MNDO. The experimental value is 1.0 kcal mol^{-1}.

7 Continue the plot of Computer Project 13-5 to twist angles of 70 and 80°. Sketch what you think the entire plot will look like for a full rotation of one methyl group relative to the other (-180 to $180°$).

8 Estimate the enthalpy change for the reaction

$$CO(g) + \tfrac{1}{2}O_2(g) = CO_2(g)$$

at 25°C from MNDO calculations. Take $\Delta H_f(O_2) = 0$. The experimental value for this enthalpy of reaction is -67.7 kcal mol^{-1}.

9 According to MNDO, which is the more stable conformer of propene, the staggered or the eclipsed? Compare the results with those obtained by molecular mechanics (Chapter 10).

10 Obtain the dipole moment of methylenecyclopropene by MNDO. Compare the results with those obtained from Huckel molecular orbital calculations (Chapter 8).

REFERENCES

Atkins, 1986. *Physical Chemistry*. Freeman, New York.

Bingham, Dewar, and Lo, 1975. *J. Amer. Chem. Soc.*, **97**, 1285.

Clark, T., 1985. *Handbook of Computational Chemistry*. Wiley, New York.

Dence, J. B. and Diestler, D. J., 1987. *Intermediate Physical Chemistry*. Wiley, New York.

Dewar, M. J. S., 1975. *Science*, **187**, 1037.

Dewar, M. J. S. and Thiel, W., 1977. *J. Amer. Chem. Soc.*, **99**, 4907.

Dewar, M. J. S., Zoebsch, E. G., Healy, E. F., and Stewart, J. J. P., 1985. *J. Amer. Chem. Soc.*, **107**, 3902.

Hertzberg, 1970. *J. Mol. Spectroscopy*, **33**, 147.

Kistiakowsky et al., 1935. *J. Amer. Chem. Soc.*, **57**, 65.

Klopman, G. and Evans, R. C., 1977. *Modern Theoretical Chemistry*, G. A. Segal, ed. Plenum, New York, Vol. 7, Chapter 2.

Mc Quarrie, D. A., 1983. *Quantum Chemistry*. University Science Books, Mill Valley, CA.

Turner, D. W., Baker, C., Baker, A. D., and Brundle, C. R., 1970. *Molecular Photoelectron Spectroscopy*. Wiley-Interscience, London.

A | *Listing of Programs and Sources*

BASIC

All BASIC programs were written by the author and are on the disk included with this book. Many contain sample data sets in the form of DATA statements. These can, of course, be altered for general use.

EDLIN

This is the line editor available as part of the standard software package with the Tandy 1000 series of computers. (Tandy/TM Tandy Corp., Fort Worth, TX 76102.) A short glossary of EDLIN commands is provided with the software.

GAUSSIAN70 (QCMPE 061)

This package is available from Quantum Chemistry Program Exchange, Chemistry Department, Room 204, Indiana University, Bloomington IN 47401. Further description and use are found in Chapter 12.

HMO (disk)

Modified and adapted for microcomputer from an original program by Greenwood published by Wiley.

MICROCADPAC

This is a simplified version of the Cambridge *ab initio* package available from S. M. Colowell and N. C. Handy, Department of Theoretical Chemistry, University Chemical Laboratories, Lensfield Road, Cambridge, CB2 1EW, UK. See also MICROMOL.

215

MICROMOL

Full *ab initio* programs (2) available from LYNXVALE, Wolfson Cambridge Industrial Unit, 20 Trumpington Street, Cambridge CB2 1QA, UK.

MM2 (QCMPE 004)

This program is a modification of Allinger's MM2 by D. P. Chong and is available from QCPE in both compiled and source code. The original mainframe program is QCPE 395.

MMX

A version of Allinger's MMP2 by J. J. Gajewski and K. E. Gilbert. This program is available from Serena Software, PO Box 3076, Bloomington, IN 47402-3076.

MNDO

This program is a modification of the original [QCPE 353 (1977)] mainframe program by Dewar and Thiel [Chapter 13, Dewar and Thiel (1977)], revised by K. E. Gilbert and J. J. Gajewski, Indiana University, 1986. It is available from Serena Software, PO Box 3076, Bloomington IN 47402-3076.

PCMIO

This is a companion program to MMX that plots the molecule after MM optimization. It also generates z-matrix input for *ab initio* and semiempirical calculations. It is not discussed here because it slightly exceeds the (minimal) hardware assumed for the exercises in this book. This program is available from Serena Software, PO Box 3076, Bloomington IN 47402-3076.

PCMODEL

Molecular Graphics for IBM PC/TM IBM Version 2.0, by J. G. Henkel and F. H. Clarke, is available from Academic Press, Inc., Harcourt Brace Jovanovich, Publishers, Orlando FL 32887.

PPP-MO (QCMPE 054)

Pariser-Parr-Pople Molecular Orbital Calculations by J. Griffiths and J. C. Lasch. This program is available from QCPE. For further description and use, see Chapter 9.

SCF (disk)

Modified and adapted from an original program by Greenwood published by Wiley.

SIGMAPLOT

Sigma-Plot Publication Quality Graphs for the Scientist version 3.10 (1988) by S. Rubenstein, development tools by J. Norby, and manual by R. Mitchell. The program and very extensive documentation are available from Jandel Scientific, 65 Koch Road, Corte Madera, CA 94925. Further description and use can be found in Chapter 11.

Index

219